Chemical Engineering and Technology

Chemical Engineering and Technology

Editor: Ian Goodwin

NY RESEARCH PRESS

New York

Published by NY Research Press
118-35 Queens Blvd., Suite 400,
Forest Hills, NY 11375, USA
www.nyresearchpress.com

Chemical Engineering and Technology
Edited by Ian Goodwin

International Standard Book Number: 978-1-63238-591-8 (Hardback)

Cataloging-in-Publication Data

Chemical engineering and technology / edited by Ian Goodwin.
p. cm.
Includes bibliographical references and index.
ISBN 978-1-63238-591-8
1. Chemical engineering. 2. Chemistry, Technical. I. Goodwin, Ian.
TP155 .C45 2018
660.2--dc23

Contents

Preface

This book elucidates new techniques of chemical technology and their applications in a multidisciplinary approach. It discusses in detail the concepts and methods used in this field. Chemical engineering refers to that sub-field of engineering, which transforms, produces and uses materials, chemicals and energy by applying laws of biochemistry, physics, microbiology and applied mathematics. The main concepts used in this field are process design, chemical reaction engineering, transport phenomena, etc. Such selected concepts that redefine chemical engineering have been presented in this text. The topics covered in it offer the readers new insights in this field. This textbook, with its detailed analyses and data, will prove immensely beneficial to professionals and students involved in this area at various levels.

A detailed account of the significant topics covered in this book is provided below:

Chapter 1- Chemical technology has directly and indirectly helped fulfill the basic necessities of life such as food, shelter and clothing. It is used in manufacture of products like petrochemicals, soap, fertilizers, sugar, paper etc. This is an introductory chapter which will introduce briefly all the significant aspects of chemical technology.

Chapter 2- Coal is a necessary component for industrial use. It is used as a fuel owing to its combustible nature. It is processed using various methods such as coal carbonization, hydrogenation of coal, coal gasification, etc. This section discusses the methods of coal processing in a critical manner providing key analysis to the subject matter.

Chapter 3- Crude oil is a multicomponent mixture that needs to be refined to obtain valuable products such as diesel oil, liquefied petroleum gas, petrol, etc. The processes used in refining include distillation, alkylation, isomerization, cracking, etc. The following chapter elucidates the varied processes and mechanisms associated with this area of study.

Chapter 4- The products obtained after petroleum distillation is called petrochemicals. The derived products include C_1, C_2, C_3 and C_4 compounds, and benzene. Petrochemicals can be classified into light petrochemicals, medium chemicals and heavy petrochemicals. The aspects elucidated in this section are of vital importance, and provide a better understanding of petrochemicals.

I would like to make a special mention of my publisher who considered me worthy of this opportunity and also supported me throughout the process. I would also like to thank the editing team at the back-end who extended their help whenever required.

Editor

Fundamentals of Chemical Technology

Chemical technology has directly and indirectly helped fulfill the basic necessities of life such as food, shelter and clothing. It is used in manufacture of products like petrochemicals, soap, fertilizers, sugar, paper etc. This is an introductory chapter which will introduce briefly all the significant aspects of chemical technology.

Chemical Industry

Chemical industry is one of the oldest industries and playing an important role in the social, cultural and economic growth of a nation and in providing basic needs of humankind - food, shelter and clothing have become an indispensable part of our life. Figure illustrate the role of chemical industry in daily life. It is one of most diversified of all industrial sectors covering thousands of products. Chemical industry includes basic chemicals and its product, petrochemicals, fertilizers, paints & varnishes, gases, soap and detergent, perfumes, pharmaceuticals and covers thousands of products, which are finding use in our daily life from industrial to household goods. Structure of organic chemical industry is shown in Figure. Various products are finding use in various fields packaging to agriculture, automobiles to telecommunication, construction to home appliances, health care to personal care, explosive, pesticides to fertilizer, textile to tire cord, chemicals to pharmaceuticals. Indian chemical industry plays an important role in the overall development of Indian economy and contributes about 3% in the GDP of the country. It comprises large scale, medium scale and small units.

The chemical industry is a key contributor to the world economy and produces more than 8000 products, which is a vital part of agricultural and industrial development in India and has key linkages with several other downstream industries such as automotive, consumer durables, engineering and food processing. Organic chemicals are one of the important sectors of the Indian chemical industry, which provide a vital development role by providing petroleum products, chemical feedstock, basic chemicals, intermediates, and important products like polymer, synthetic fibre, synthetic rubber, paints, varnish, pesticides and explosives, dyes, specialty chemicals. Major feedstocks for chemical industries are coal, petroleum, biomass, oils and fats, sulphur, salt, lime stone, rock phosphate etc.

Table : Major Products of Chemical Industries and their Area of Application

Group of Product	Areas
Plastics and Polymers	Agricultural water management, packaging, automobiles, telecommunications, health and hygiene, education
Synthetic rubber	Transportation Industry, Textile, Industrial equipment lining
Synthetic fiber	Non-oven and woven fibre in automobile , hosiery, textile
Soap and Synthetic detergents	Health and hygiene domestic as well as industrial
Industrial chemicals	Drugs & pharmaceuticals, pesticides, explosives, surface loading, dyes, lube additives, adhesive oil field, antioxidants, chemicals, metal extraction, printing ink, paints
Sugar & Alcohol	Food, Alcoholic Brewages, Chemical Feed Stock, Ethoxylate, biofuel
Pulp & Paper	Writing & Printing Paper, Culture Paper, News Printing Paper, Tissue Paper, Packaging Paper
Fertiliser	Agriculture, Chemical industry(ammonia and urea)
Agrochemicals	Pesticides
Mineral acids	Chemical industry- organic and inorganic

Chemical Industry
Food
Fertilizer & Agrochemical
Clothing
Synthetic fibers, Dyestuffs, Textiles, Auxiliaries, Specialty Chemicals
Shelter
Polymer composites, Coating, New Performance Materials
Health Care
Pharmaceuticals, Polymers, Synthetics, Detergent
Quality of Life
Transportation, Education, Fuel, Electricity, Energy, Water supply, Management, Communication, Polymers & Industrial Chemicals

Figure: Role of Chemical Industry

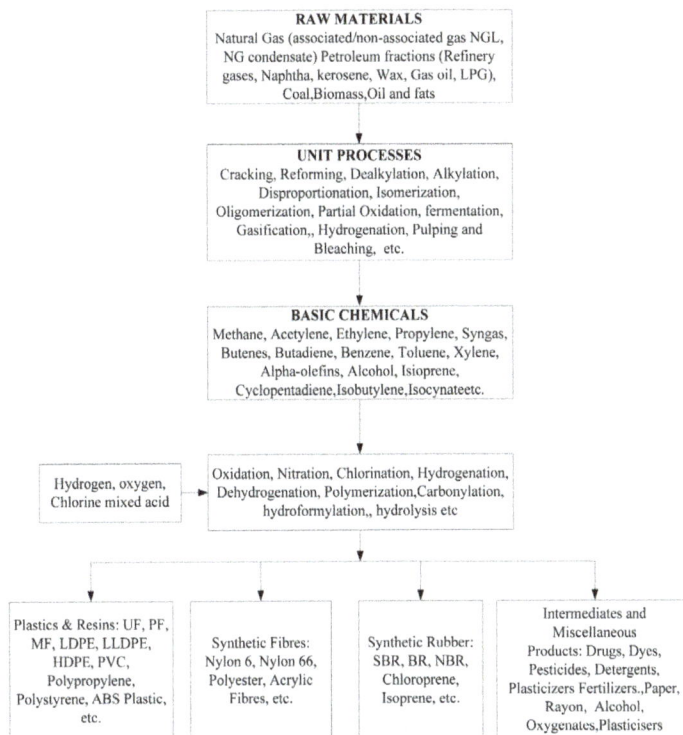

Structure of Organic Chemical Industry

Chemical Industry and Technological Development in Chemical Industry

Chemical process industry has evolved considerably over the last century largely in response to changing societal requirements and changing raw material availability and environmental issues. Some of the major technological development in chemical industry are

- Leblanck process to Solvay and modified Solvay process

- Lead chamber to Contact process (single absorption and DDCA

- Diaphragm process to Mercury and Mercury to Membrane

- Wet to Dry cement Process

- Coal chemicals to alcohol based chemicals to petroleum based chemicals and vice versa

- Acetylene based chemicals to other routes

- Claus to Super Claus process

- Wood based paper to agro-based and waste paper based

- Pulping to biopulping
- Stone ground wood pulping to thermo mechanical pulping
- Chlorine to oxygen bleaching and enzymatic bleaching
- Sulphur to pyrite based sulphuric acid plant
- Conventional aluminium and iron based catalyst to zeolite based catalyst
- Coal based Fertilizer to Natural gas and Naphtha based Fertilizers
- Coal and Alcohol based chemical to Petroleum based chemicals
- Thermal cracking to Catalytic cracking
- FCC to Deep catalytic cracking for olefin and Hydrocracking for processing heavier crude
- Naphtha reforming to isomerisation
- Acid catalyst to solid acid catalyst in alkylation process
- Naphtha steam cracking to gas cracking
- Conventional petroleum fuel to biofuel
- Coal as fuel to coal as chemical
- Coal gasification to Petrocoke and Biomass gasification
- Chemical pesticide to bio pesticide
- Chemical fertilise to biofertliser
- Soap to detergent, Non biodegradable detergent to Biodegradable detergent
- Natural gas to coal bed methane, shale gas, gas hydrate
- Dimethyl terephthalate(DMT) to Purified terephthalic acid(PTA)
- Conventional caprolactam to ammonium sulphate free caprolactam
- Natural fibre to Synthetic Fibre
- Natural rubber to synthetic rubber
- Petroleum refinery to Natural gas refinery and Biorefinery
- Petroleum refinery to Petrochemical refinery
- Conventional gasification to underground gasification

- Conventional drilling to horizontal drilling and hydrofracturing

- Gas to liquid and Methanol to olefin technology

- Coal to methanol and olefin

- Conventional desulphurization to ultra desulphurization processes and bio-desulphuristion

- Polymer to b opolymer

- Conventional Ziegler Natta catalyst to metallocene catalyst

Needed of Chemical Process Technology

A process engineer at operation in chemical plant shall have a deeper understanding of the technology on which the process plant is built to produce the profit making chemicals. A chemical engineer with sound knowledge in process technology has the following distinct advantages:

a) Ability to clearly distinguish the functional role and importance of various processes and operations in the process plan

b) Technical knowledge with respect to the selection of important parameters such as Temperature, Pressure and underlying physical principles of a process.

c) Ability to distinguish various process streams and their conditions of operation (Temperature, pressure and phases)

d) Basic knowledge for process troubleshooting and necessary safety precautions associated to a process/operation.

How to Master Fundamentals of Process Technology?

To master chemical process technology five crucial steps are involved namely:

a) Raw-Materials and reactions: A chosen process route to manufacture desired chemicals with appropriate purities will eventually lead to preparing a list of raw-materials and utilities. Thereby, prominent reactions can be also known.

b) Conceptual process flow-sheet: A conceptual process flow-sheet where a chemical engineer has an abstract representation of the actual process flow-sheet will enable quicker learning. A conceptual process flow-sheet typically constitute the following attributes:

- Raw-material purification (Solid-fluid operations such as cyclone separators, ...bag filters etc.)

- Raw-material processing (Heat exchange operations such as furnace ...heating, cooling etc.)

- Raw-material to product transformation (Reaction operations using CSTR, ...PFR, PBR and Batch reactors)

- Product purification (In separation processes such as flash, distillation, ...absorption and extraction)

- Product processing (heat exchangers, phase change units)

- Recycle of un-reacted raw-materials as recycle streams to the reaction ...operations.

- Technological know-how of each unit operation/process to eventually identify its functional role in the process flow-sheet

- Temperature, pressure and composition of streams entering the unit ...process/ operation

- Temperature, pressure and composition of streams leaving the unit ...process/ operation

- Phases of various streams entering and leaving the unit process/operation

- Associated waste and environmental hazardous products generation

c) Process intensification in the form of heat-integration, stream utilization and waste reduction and multiple recycle streams: These options are in fact optional and they enrich the energy enhancement and waste reduction efficiency of a process plant. Originally, chemical plants developed without such process intensification policies have been subjected to rigorous research and case study investigations to identify opportunities for cost reduction and better energy/waste management.

d) Additional critical issues related to various unit operations/processes

- Cost reduction approaches/process economics: How the operating variables (such as temperature, pressure, flow-rates, reflux rates, heating rate etc.) affect the cost of the unit operation/process

- Safety issues: What safety issues are most relevant and need frequent ...monitoring

e) Alternate technologies: For a desired function of a process unit, can thereby alternate technologies that could reduce the cost and even then provide the same functional role and desired flow rates and compositions of the emanating streams.

How Much to Memorize for a Chosen Technology ?

To a large extent, University education expects a chemical engineering undergraduate student to remember and draw at least a conceptual flow-sheet. However, when a systematic approach is not adopted in the learning process, it is rather difficult to remember all flow-sheets and relate to the logic behind their role in the process topology.

Therefore, a well-trained student in process technology remembers process flow-sheets with logical sequence of unit processes/operations and not by strict memory.

Advantages of Suggested Learning Approach for Mastering Process Technology

a) Trains student to be more analytical/concept-oriented rather than with memorized knowledge that is bereft of logical reasoning

b) Systematic approach enables the growth of students' interest in the subject.

c) Additional concepts further reveal to the student how to gradually complicate process technologies for maximum efficiency.

d) Inculcate strong interest in the student towards technology research and innovation by enabling a learning environment that fundamentally targets the technological know-how.

Prominent Unit-operations and Unit-processes in Chemical Industry

A detailed summary of various prominent unit operations/processes and their functional role in the chemical plant are summarized in Table along with suitable figures.

Category	Unit operations/processes	Functional role
fluid operations	a) Centrifugal pump b) Reciprocating pump c) Compressor d) Expander	a) To pressurize liquids and gases. b) To depressurize gases
solid operations	a) Crusher b) Grinder	a) To reduce the size of solids
Solid-fluid separators	a) Cyclone separator b) Centrifuge c) Electrostatic precipitator d) Classifier & Thickener e) Liquid-liquid separator	a) To separate solid particles from solid-liquid/gas mixtures
Heat exchangers	a) Shell & Tube heat exchangers b) Fired heaters and furnaces c) Coolers	a) To either remove or add heat to process streams so as to meet desired conditions in other units. b) Either utilities or other process streams are used to carry out heating/cooling requirements.
Mass transfer units	a) Phase separation b) Distillation c) Absorption d) Stripping e) Adsorption f) Extraction g) Leaching h) Crystallization i) Membrane	a) To separate a feed into products with different compositions. b) A third agent (heat or compound) is usually used to carry out separation.

Reactor units	a) Completely stirred tank reactor (CSTR) b) Plug flow reactor (PFR) c) Packed bed reactors (PBR) d) Slurry & Trickle bed reactors	a) To carry out reactions in homogenous fluids (gases/liquids). b) To carry out catalytic and multi-phase reactions.

Table: Important unit operations/unit processes and their functional role in chemical process technology.

A pictorial representation of various unit processes and operations that are often encountered in chemical process flowsheets is presented in Table. Along with these figures, their function role in the process technology is also presented.

Process Technology	Functional Role
Reactors a) CSTR b) Batch-Reactor c) PFR d) Packed bed reactor e) Trickle bed reactor f) Fluidized bed reactor	- Central and most important process technology in process flow - Carry out desired reactive transformations
a) CSTR CSTR	- Well mixed reaction system - Homogenous liquid/gas phase reaction - Most easy configuration - Temperature control through Jacket - Reactant instantaneously reaches lowest concentration -Mostinexpensive to design and operate
b) Batch Reactor 	- Has a simple design, with the requirement of very little supporting equipments - Ideal for small scale experimental studies on reactor kinetics - Can be used industrially for treatment of very small quantities of materials.

c) PFR	- Homogenous liquid/gas phase reaction
	- Reactant gradually reaches low concentrations
	- Good control over temperature
	- Temperature control through jacket (not shown)
d) Packed Bed Reactor (PBR)	- Hetero------geneous reaction
	- Packing to act as catalyst
	- Packing packed in tubes
	- Shell fed with cooling/heating fluid (optional)
	- Most common in process flow sheets

e) Trickle Bed Reactor	- Multi-phase reaction
	- If the reaction is not catalytic, packing serves to enhance interfacial area
	- If the reaction is catalytic, packing acts as a catalyst as well
	- Complicated design
f) Fluidized bed reactor	- Provides highest mass, heat and hence reaction rates for solid-fluid reactions
	- Very commonly deployed in petroleum refineries (catalytic cracking)
	- Complicated accessories (shown) and control system required
	- The accessories are for catalyst re-generation and transport.

Process Technology	Functional Role
Separators: a) Batch distillation b) Continuous distillation c) Absorption d) Stripping e) Liquid-liquid extraction f) Leaching g) Crystallization h) Drying i) Flash separator j) Membrane separator k) Packed bed contactor	- Most important process technology - Provides desired separation be-tween phases and streams - Located next to the reactor as 100 % conversions are very rare in indus-trial practice
a) Batch distillation column 	- Used to separate a liquid mixture based on relative volatility (differ-ences in boiling points) - Operated in batch mode
b) Continuous distillation (Fractionator) 	- The most important separation technology in process flow sheets - Provides very pure products - Differences in boiling points is the working principle - Energy intensive operation - Usually multi-component distillation is apparent in industrial practice

c) Absorption column Absorber used for taking up a soluble gas in a solvent liquid.	- Used to absorb components from gaseous stream - Solvent is used - Usually followed with stripper to re-generate the fresh solvent - Operated at low temperature and moderate/high pressure
d) Stripper Stripper used for removing a soluble gas from solution by counter current contact with an inert gas.	- Steam/Hot gas is used to strip the gas - Regenerated solvent used for absorption
e) Liquid Liquid extraction 	- Used to separate components from a liquid with a liquid solvent - Consists of a series of mixers and sepa-rators - Produces extract (rich with solvent and components extracted) and raf-finate (product with lean extractants)

f) Leaching	- A liquid solvent extracts components from a solid
Feed slurry Clear liquid · · · Clear liquid Free settling zone Transition zone Thickened compression zone Revolving rake Thickened sludge	- A liquid solvent extracts components from a solid - High interfacial area between solid/liquid is required to enhance extraction capability

g) Crystallization	- Used to crystallize solids from a slurry/ super-saturated solution
Vapor Steam Condensate Heater Feed Crystals Product	- Used to crystallize solids from a slurry/ super-saturated solution - Fine crystals added to serve as nucleating agent

H) Spray Drier	
Liquid feed Hot air Droplets Spray chamber Solid Dry product	- Liquid slurry is sprayed in the form of droplets - Hot gas (air) dries the solid - Enables very good control over the product particle size

i) Rotary drier	- Through rotation, an agitated liquid film is dried to obtain the dried solid.
j) Flash separator	- Very common technology to separate liquid streams at high pressure and lower temperatures. - Upon pressure reduction/ heating, low boiling components separate as vapor phase and yield a liquid phase. - Complete separation only possible for fewer components
k) Membrane separation	- A semi-permeable barrier (membrane) is used to separate feed streams based on concentration difference/ pressure difference. - Various types of processes available - New technology in process industries.

l) Packed bed contactor	- Used for absorption/stripping operations
	- Packing serves to enhance gas/liquid interfacial area

Heat exchange equipment a) Shell & Tube heat exchanger b) Fired heater c) Multiple effect evaporator d) Quenching	- Very prominent equipment to heat/cool process fluids - Include steam/power generation as well!
a) Shell & Tube heat exchanger 	- Most common equipment in process industries - Tube fed with a fluid and shell is fed with another fluid - Process heat is transferred across the tube - No mixing of tube fluid and shell fluid allowed
b) Fired heater 	- Used to heat streams to extremely high temperatures - High temperatures generated by burning fuel oil/fuel gas - Complicated design for maximum heat transfer efficiency - Shell & tube type/radiation type designs usually adopted

c) Multiple effect evaporator	- Common equipment to concentrate a solid-liquid stream from low concentration to high concentrations. - Steam utility is optimized by adopting process intensification method.
d) Quenching	- Direct heat transfer equipment - Involves cooling/heating a fluid with direct contact with a stream - Commonly used for streams emanating with very high temperatures from reactions/furnaces.
Solid-fluid process technology a) Cyclone separator b) Centrifuge c) Electrostatic separator d) Thickener e) Liquid- liquid separator Filter press	- Used for separating solids from solid-liquid or solid-gas mixtures
a) Cyclone separator	- Separates fine solid particles from a gas-solid mixture - Uses centrifugation as working principle. - Very good separation of solid and liquid possible

b) Centrifuge Liquid Slurry Solids Ex: Sugar is extracted from sugar slurry.	- Separates solids from solid-liquid mixture - Uses the principle of centrifugation for separation - Very good separation of solid and liquid possible
c) Electrostatic separator Discharge insulator Collecting(+ve) plate Discharge (+ve) electrons Collecting(+ve) plate DC output Gas inlet Clean gas AC input Grounding	- Separates solids from solid-liquid mixture - Uses the principle of charged surfaces to separate the solids - Very common in process technologies.
d) Thickener Feed slurry Clear liquid Clear liquid Free settling zone Transition zone Thickened compression zone Revolving rake Thickened sludge	- Separates a slurry (solid-liquid) into a sludge and clarified liquid - Settling is adopted as working principle.

e) Liquid-liquid separator Liquid A → Solvent → → Extract (Solvent + A) → Raffinate (Lean A)	- Uses decantation as working principle based on density difference.
f) Filter Overflow Solids Filtrate discharge ← Filter cloth → Slurry feed	- Separates a solid from solid-fluid mixture - Uses a fine mesh/cloth to separate under pressure.
Fluid transport a) Centrifugal pump b) Reciprocating pump c) Steam jet ejector d) Compressor e) Expander	- Very important to achieve process conditions desired in other important processes such as reactors and separators.
a) Centrifugal pump Suction	- Energizes liquids to moderately high pressure.
b) Reciprocating pump Discharge Suction	- Energizes liquids to very high pressures.

c) Steam jet	- Used for providing vacuum (low pressure) in various units
Steam inlet Discharge Diffuser Steam nozzle Vapor inlet	- Common in process flow sheets.
d) Compressor	- Enhances pressure of gases to high values.
Gas at low pressure → → Gas at high pressure	
e) Expander	- Reduces pressure of gas to lower values
Gas at high pressure → → Gas at low pressure Shaft work Electricity	- Recovered energy used for shaft work or power generation (electricity).

Size Reducer: a) Crusher b) Grinder	- Used for reducing size of solids and enhance their surface area to facilitate higher mass transfer and reaction rates.
a) Crusher Feed Product	- Continuous operation - Size control is very easy

b) Grinder	- Batch operation
	- Achieving size control is difficult.

Storage: a) Storage tank b) Pressurized spheres	- Used to store fluids and gases.
a) Storage tank 	- Used especially for liquid fuels
b) Pressurized spheres 	- Used to store gaseous fuels.
Table: Summary of various prominent process technologies and their functional role in process flow sheets.	

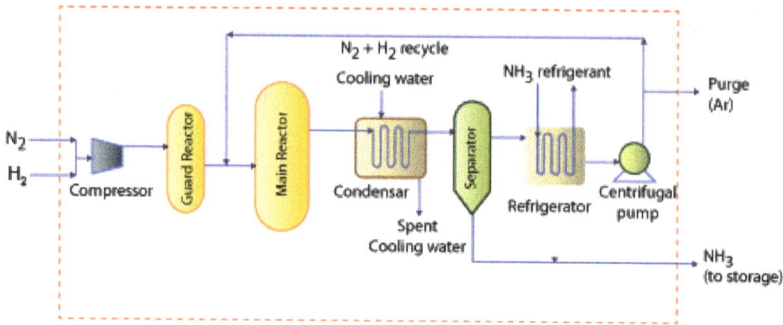

Process flow sheet for ammonia manufacture using Haber's process.

Illustration of Quickly Learning Process Technology: Ammonia Manufacture using Haber's Process

a) Reaction: The Haber process combines nitrogen from the air with hydrogen derived mainly from natural gas (methane) into ammonia. The reaction is reversible and the production of ammonia is exothermic. Nitrogen and hydrogen will not react under normal circumstances. Special conditions are required for them to react together at a decent rate forming a decent yield of ammonia. These conditions are T = 400°C; P = 200 atm; Iron catalyst with KOH promoter.

b) Raw materials: H_2 from synthesis gas, N_2 from synthesis gas/air liquefaction process

c) Process technology: Illustrated in Figure

d) Unit processes: Feed Guard Converter; Main Reactor

e) Unit operations in the technology: Condensation/Gas Liquefaction; Separation; Refrigeration; Centrifugal Re-circulator

f) Striking feature: Conversions are low (8 - 30 % per pass) and hence large recycle flow rates exist.

g) Functional role of various processes

a. Feed guard converter:

» CO and CO_2 conversion to CH_4 and removal of traces of H_2O, H_2S, P and Arsenic.

» These compounds could interfere with the main haber's reaction as well as poison the catalyst.

b. Main reactor:

» Cold reactants enter reactor from reactor bottom and outer periphery to absorb heat generated in the reversible reaction.

» Carbon steel used for thick wall pressure vessel and internal tubes

» Gas phase reaction at $500 - 550°$ C and $100 - 200$ atms.

» Pre-heated gas flows through the tube inside with porous iron catalyst at reaction conditions.

» Catalyst removed from the converters is re-fused in an electric furnace.

c. Condenser:

» Complete liquefaction not possible due to vapor liquid equilibrium between NH_3 in vapor and liquid phases.

» Cooling fluid: Chilled water

» Incoming stream has: Gaseous NH_3, un reacted N_2 and H_2, impurity gases like CO, CO_2, CH_4

» Product stream has: Liquefied NH_3, Vapor phase NH_3 in equilibrium with liquid, non-condensable gases such as N_2 and H_2

» System pressure is high therefore NH_3 bound to be present in higher concentrations in the vapor.

d. Separator:

» Working principle: Density difference between vapor and liquid

» Liquefied NH_3 is separated from the un-reacted gases (NH_3 still present in the vapors).

e. Refrigeration:

» Why?: NH_3 available in vapors needs to be condensed. Therefore, refrigeration is required for gaseous product emanating from the separator.

» The condensed NH3 emanates at -15°C.

» After refrigeration, the un-reacted N_2 and H_2 are recycled to the reactor.

f. Centrifugal pump:

» A centrifugal pump adjusts the pressure of the stream from the separator to the pressure of the feed entering the reactor

» A purge stream exists to facilitate the removal of constitutes such as Argon.

g. Striking feature:

All units such as Condenser, Separator, Refrigerator operate at high pressure. This is because loosing pressure is not at all beneficial as the un-reacted reactants (corresponding to large quantity in this case due to low conversion in the reversible reaction) need to be supplied back to the reactor at the reactor inlet pressure conditions.

Elementary processes

Chemical processes usually have three interrelated elementary processes

- Transfer of reactants to the reaction zone

- Chemical reactions involving various unit processes

- Separation of the products from the reaction zone using various unit operations

Processes may involve homogeneous system or heterogeneous systems. In homogeneous system, reactants are in same phase-liquid, gases or solids while heterogeneous system include two or more phases; gas liquid, gas–solid, gas-gas, liquid–liquid, liquid solid etc. Various type reactions involve maybe reversible or irreversible, endothermic or exothermic, catalytic or non-catalytic. Various variables affecting chemical reactions are temperature pressure, composition, catalyst activity, catalyst selectivity, catalyst stability, catalyst life, the rate of heat and mass transfer. The reaction may be carried out in batch, semi batch or continuous. Reactors may be batch, plug flow, CSTR. It may be isothermal or adiabatic. Catalytic reactors may be packed bed, moving bed or fluidised bed Along with knowledge of various unit processes and unit operation following information are very important for the development of a process and its commercialization [Austin,1984] Basic Chemical data: Yield conversion, kinetics

- Material and energy balance, raw material and energy consumption per tone of product, energy changes

- Batch vs Continuous, process flow diagram

- Chemical process selection: design and operation, pilot plant data, Equipment required, material of construction

- Chemical Process Control and Instrumentation

- Chemical Process Economics: Competing processes, Material and, Energy cost, Labour, Overall Cost of production

- Market evaluation: Purity of product and uniformity of product for further processing

- Plant Location

- Environment, Health, Safety and Hazard

- Construction, Erection and Commissioning

- Management for Productivity and creativity: Training of plant personals and motivation at all levels

- Research, Development and patent

- Process Intensification

  Inorder to improve productivity and make the process cost effective and for improving overall economy, compact , safe, energy efficient and environmentally sustainable plant, process intensification has become very important and industry is looking beyond the traditional chemical engineering.

Unit Processes and unit Operations in Chemical Process Industries

Chemical process is combination of unit processes and Unit operation. Unit process involves principle chemical conversions leading to synthesis of various useful product and provide basic information regarding the reaction temperature and pressure, extent of chemical conversions and yield of product of reaction nature of reaction whether endothermic or exothermic, type of catalyst used. Unit operations involve the physical separation of the products obtained during various unit processes. Various unit processes in chemical industries are given in Table. Various chemical reactions and its application in process industries are given in Table.

Nitration

Nitration involves the introduction of one or more nitro groups into reacting molecules using various nitrating agents like fuming, concentrated, aqueous nitric acid mixture of nitric acid and sulphuric acid in batch or continuous process. Nitration products find wide application in chemical industry as solvent, dyestuff, pharmaceuticals, explosive, chemical intermediates. Typical products: TNT, Nitrobenzene, m-dinitrobenzene, nitroacetanilide, alpha nitronaphthalene, nitroparaffins

Table: Unit Processes in Chemical Process Industries

Alkylation and Hydro delkylation	Decomposition
Acylation	Fermentation
Ammonoxidation	Halogenation
Amination by reduction	Hydsogenation
Amination	Hydrohenatlysis
Aromatisation	Hydroformylation
Amination by ammonalysis	Hydro lysis
Calcination	Hydration

Carbonation	Hydroammonalysis
Causticisation	Isomerisation
Chlorination and Oxy chlorination	Neutralistion
Condensation	Nitration
Biomethhanation	Methanation
Carbinisation	
Disproportination	Oxidation and partial oxidation
Cracking; Thermal, steam cracking, catalytic cracking	Pyrolysis
Dehydration	Polymeristion: Addition and condensation Chain growth and Step growth,Bulk, Emulsion, suspension, solution, Radical and coordination polymeristion
Dehydrogenation	Reduction
Ditozitation and coupling	Reforming: Steam reforming Catalytic reforming
Gasification of coal and biomass	Sulphidation
Desulphurisation and hydro desulphurisation	Sulphonatiomn
Electrolysis	Sulphation
Etherification	Xanthation
Estertification and Trans Estrerificartion	

Table : Important Chemical Reaction and their Application in Chemical Process Industries

Reaction	Description
Fisher-Tropsch (FT) Process	The Fisher-Tropsch process produce a variety of hydrocarbons (alkanes: $C_nH_{(2n+2)}$) by involves a series of chemical reaction. $(2n+1) H_2 + nCO \rightarrow C_nH_{(2n+2)} + nH_2O$ FT process is used for synthesis of alkanes.
Friedel-Crafts reactions	In this reaction attach substituent's to an aromatic ring. Two main types of Friedel-Crafts reaction are acylations reaction and alkylation reactions, both proceeding by electrophilic aromatic substitution. Friedel-Crafts process used in alkelation reactions.
Oxosynthesis Reactions	In this process Isomeric mixture of normal- and iso-aldehydes get produces by utilizing syngas (CO and H_2) and olefinic hydrocarbons as reactants. It is exothermic process, this process thermodynamically favorable at ambient pressure and temperatures. This reaction also called as hydroformylation reaction. $RCH=CH_2+CO+H_2 \rightarrow RCH_2CH_2CHO + R(CH_3)CHCHO$ Oxosynthesis used for production of alcohols.

Hofman Process	In this process, organic reaction of primary amide converts into a primary amine with one fewer carbon atom. Hofman process typical examples are conversion of aliphatic amides to aliphatic amines and aromatic amides to aromatic amines.
Free-Radical Reaction	Any chemical reaction involving free radicals, generally radical generated from radical initiators such as peroxide or azo bis compounds. Radical reactions are chain reactions with chain initiation, propagation and termination steps. Free radical reactions are used many organic synthesis and polymerization reactions
Beckmann rearrangement	Beckmann rearrangement is acid catalyzed rearrangement of an oxime to an amide, which developed by German chemist Ernst Otto. A typical example of Beckmann rearrangement is synthesis of caprolactam from cyclohexanone. Caprolactam is monomer for nylon 6.
Wackers Process	Wacker process is similar to hydroformylation and used for aldehyde compounds. A typical example of wacker process is oxidation of ethylene to acetaldehyde in the presence of Pd catalyst. $[PdCl_4]^-+C_2H_4+H_2O \rightarrow CH_3CHO+Pd+2HCl+2Cl^-$ $Pd^{++} CuCl_2+2Cl^- \rightarrow [PdCl4]^-+2CuCl$ $2CuCl+ 0.5 O_2+2HCl \rightarrow 2CuCl_2+H_2O$

Example Preparation of TNT (trinitrotoluene)

TNT is produce in a three-step process: First, toluene is nitrated with a mixture of sulfuric acid and nitric acid to produce mono-nitrotoluene or MNT. The MNT is separated and then renitrated to dinitrotoluene or DNT. In the final step, the DNT is nitrated to trinitrotoluene or TNT using an anhydrous mixture of nitric acid and oleum.

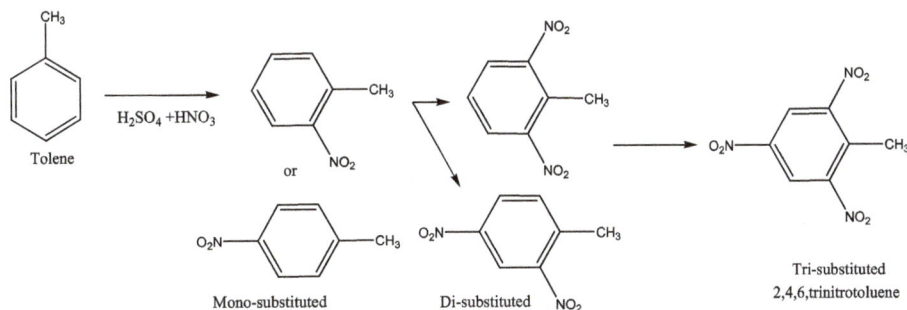

Halogenation

Halogens involve introduction of one or more halogen groups into a organic compound for making various chlorine, bromine, iodine, fluorine organic derivatives. All though chlorine derivatives find larger application, however some of the bromine and fluorine derivatives are also important. Various chlorinating agents are chlorine, HCl, phosgene sulfuric chloride, hypochlorite, bromination, bromine, hydrobromic acid, bromide, bromated, alkaline hypobromites. In iodination iodine, hydroiodic acid and alkali hypoiodites

Example

Typical important chemicals are chlorinated products: Ethylene dichloride, chlorinated methanes Chloroform, Carbon tetra chloride etc) Chlorinate ethane, Chloro propane, chloro butanes,

chloroparaffins, chlorination of acetaldehyde (Chloral), alkyl halhides, Chlorobenzene, Ethylene diiodide, Chloroflurocarbon(CFCs).

Preparation of chloroform and chloroflurocarbon (CFCs)

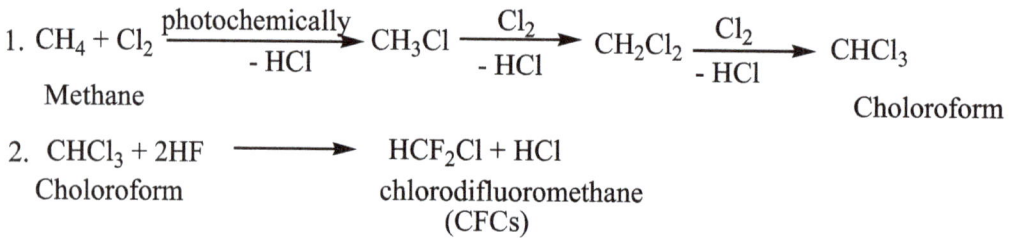

1. $CH_4 + Cl_2 \xrightarrow[\text{- HCl}]{\text{photochemically}} CH_3Cl \xrightarrow[\text{- HCl}]{Cl_2} CH_2Cl_2 \xrightarrow[\text{- HCl}]{Cl_2} CHCl_3$

    Methane   Choloroform

2. $CHCl_3 + 2HF \longrightarrow HCF_2Cl + HCl$

    Choloroform   chlorodifluoromethane
                                         (CFCs)

Sulphonation and Sulphation

Sulphonation involves the introduction of sulphonic acid group or corresponding salt like sulphonyl halide into a organic compound while sulphationinvolves introduction of $-OSO_2OH$ or $-SO_4-$. Various sulphonating agents are sulphur trioxide and compounds, sulphurdixide, sulphoalkylating agents. Some of the sulphaming agents are sulphamic acid. Apart from sulfonation and sulphamate sulpho chlorinated, sulfoxidation is also used.

Typical application of sulphonation and sulphation are production of lingo sulphonates, linear alkyl benzene sulphonate, Toluene sulphonates, phenolic sulphonates, chlorosulphonicacd, sulphamates for production of herbicide, sweetening agent (sidiumcyclohexysulphamate). Oil soluble sulphonate, saccharin.

Preparation of Saccharin

The industrial synthesis entails the reaction of hydrogen chloride with a solution of sulfur trioxide in sulfuric acid. Sulfonation by chlorosulfonic acid gives the ortho and para substituted chlorosulfones. The ortho isomer is separated and converted to the sulfonamide with ammonia. Oxidation of the methyl substituent gives the carboxylic acid, which cyclicizes to give saccharin.

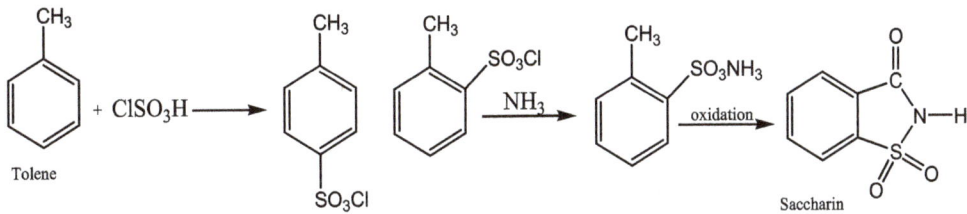

$$HCl + SO_3 \rightarrow ClSO_3H$$

Oxidation

Oxidation used extensively in the organic chemical industry for the manufacture of a large number of chemicals. Oxidation using oxygen, are combinations of various reactions like oxidation via dehydrogenation using oxygen, dehydrogenation and the introduction of oxygen and destruction of carbon, partial oxidation, peroxidation, oxidation in presence of strong oxidizing agent like $KMnO_4$, chlorate, dichromate, peroxides H_2O_2, PbO_2, MnO_2; nitric acid and nitrogen tertra oxide, oleum, ozone. Some of the important product of oxidation are aldehyde, ketone, benzyl alcohol, phthalic anhydride, ethylene oxide, vanillin, bezaldehyde, acetic acid, cumene, synthesis gas from hydrocarbon,, propylene oxide, benzoic acid, maleic acid, benzaldehyde, phtathalic anhydride. Oxidation maybe carried out either in liquid phase or vapour phase.

Preparation of Synthesis Gas from Hydrocarbon

By using the Fischer–Tropsch process, or Fischer–Tropsch synthesis, is a collection of chemical reactions that converts a mixture of carbon monoxide and hydrogen into liquid hydrocarbons.

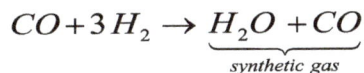

$$H_2O + CH_4 \rightarrow CO + 3H_2$$

$$CO + 3H_2 \rightarrow \underbrace{H_2O + CO}_{synthetic\ gas}$$

Hydrolysis

Generic mechanism for a hydrolysis reaction. (The 2-way yield symbol indicates an equilibrium in which hydrolysis and condensation can go both ways.)

Hydrolysis usually means the cleavage of chemical bonds by the addition of water. When a carbohydrate is broken into its component sugar molecules by hydrolysis (e.g. sucrose being broken down into glucose and fructose), this is termed saccharification. Generally, hydrolysis or saccharification is a step in the degradation of a substance OR in the language of chemistry "The reaction of cation and anion or both with water molecule due to which pH is altered, cleavage of H-O bond in hydrolysis takes place."

Hydrolysis can be the reverse of a condensation reaction in which two molecules join together into a larger one and eject a water molecule. Thus hydrolysis adds water to break down, whereas condensation builds up by removing water.

Types

Usually hydrolysis is a chemical process in which a molecule of water is added to a substance. Sometimes this addition causes both substance and water molecule to split into two parts. In such reactions, one fragment of the target molecule (or parent molecule) gains a hydrogen ion.

Salts

A common kind of hydrolysis occurs when a salt of a weak acid or weak base (or both) is dissolved in water. Water spontaneously ionizes into hydroxide anions and hydronium cations. The salt also dissociates into its constituent anions and cations. For example, sodium acetate dissociates in water into sodium and acetate ions. Sodium ions react very little with the hydroxide ions whereas the acetate ions combine with hydronium ions to produce acetic acid. In this case the net result is a relative excess of hydroxide ions, yielding a basic solution.

Strong acids also undergo hydrolysis. For example, dissolving sulfuric acid (H_2SO_4) in water is accompanied by hydrolysis to give hydronium and bisulfate, the sulfuric acid's conjugate base.

Esters and Amides

Acid–base-catalysed hydrolyses are very common; one example is the hydrolysis of amides or esters. Their hydrolysis occurs when the nucleophile (a nucleus-seeking agent, e.g., water or hydroxyl ion) attacks the carbon of the carbonyl group of the ester or amide. In an aqueous base, hydroxyl ions are better nucleophiles than polar molecules such as water. In acids, the carbonyl group becomes protonated, and this leads to a much easier nucleophilic attack. The products for both hydrolyses are compounds with carboxylic acid groups.

Perhaps the oldest commercially practiced example of ester hydrolysis is saponification (formation of soap). It is the hydrolysis of a triglyceride (fat) with an aqueous base such as sodium hydroxide (NaOH). During the process, glycerol is formed, and the fatty acids react with the base, converting them to salts. These salts are called soaps, commonly used in households.

In addition, in living systems, most biochemical reactions (including ATP hydrolysis) take place during the catalysis of enzymes. The catalytic action of enzymes allows the hydrolysis of proteins, fats, oils, and carbohydrates. As an example, one may consider proteases (enzymes that aid digestion by causing hydrolysis of peptide bonds in proteins). They catalyse the hydrolysis of interior peptide bonds in peptide chains, as opposed to exopeptidases (another class of enzymes, that catalyse the hydrolysis of terminal peptide bonds, liberating one free amino acid at a time).

However, proteases do not catalyse the hydrolysis of all kinds of proteins. Their action is stereo-selective: Only proteins with a certain tertiary structure are targeted as some kind of orienting force is needed to place the amide group in the proper position for catalysis. The necessary contacts between an enzyme and its substrates (proteins) are created because the enzyme folds in such a way as to form a crevice into which the substrate fits; the crevice also contains the catalytic groups. Therefore, proteins that do not fit into the crevice will not undergo hydrolysis. This specificity preserves the integrity of other proteins such as hormones, and therefore the biological system continues to function normally.

Upon hydrolysis, an amide converts into a carboxylic acid and an amine or ammonia (which in the presence of acid are immediately converted to ammonium salts). One of the two oxygen groups on the carboxylic acid are derived from a water molecule and the amine (or ammonia) gains the hydrogen ion. The hydrolysis of peptides gives amino acids.

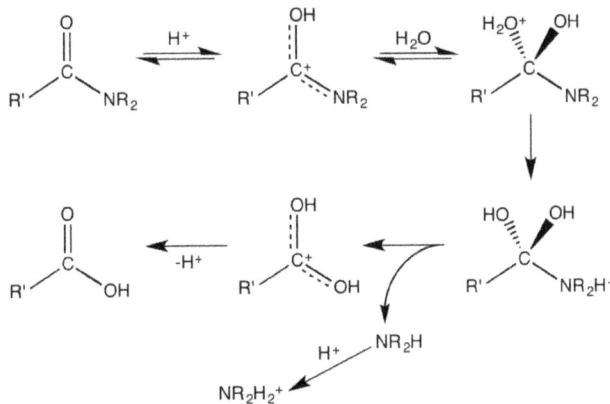

Mechanism for acid-catalyzed hydrolysis of an amide

Many polyamide polymers such as nylon 6,6 hydrolyse in the presence of strong acids. The process leads to depolymerization. For this reason nylon products fail by fracturing when exposed to small amounts of acidic water. Polyesters are also susceptible to similar polymer degradation reactions. The problem is known as environmental stress cracking.

ATP

Hydrolysis is related to energy metabolism and storage. All living cells require a continual supply of energy for two main purposes: the biosynthesis of micro and macro-molecules, and the active transport of ions and molecules across cell membranes. The energy derived from the oxidation of nutrients is not used directly but, by means of a complex and long sequence of reactions, it is channelled into a special energy-storage molecule, adenosine triphosphate (ATP). The ATP molecule contains pyrophosphate linkages (bonds formed when two phosphate units are combined together) that release energy when needed. ATP can undergo hydrolysis in two ways: the removal of terminal phosphate to form adenosine diphosphate (ADP) and inorganic phosphate, or the removal of a terminal diphosphate to yield adenosine monophosphate (AMP) and pyrophosphate. The latter usually undergoes further cleavage into its two constituent phosphates. This results in biosynthesis reactions, which usually occur in chains, that can be driven in the direction of synthesis when the phosphate bonds have undergone hydrolysis.

Polysaccharides

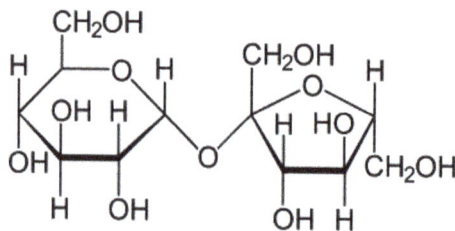

Sucrose. The glycoside bond is represented by the central oxygen atom, which holds the two monosaccharide units together

Monosaccharides can be linked together by glycosidic bonds, which can be cleaved by hydrolysis. Two, three, several or many monosaccharides thus linked form disaccharides, trisaccharides, oligosaccharides or polysaccharides, respectively. Enzymes that hydrolyse glycosidic bonds are called "glycoside hydrolases" or "glycosidases".

The best-known disaccharide is sucrose (table sugar). Hydrolysis of sucrose yields glucose and fructose. Invertase is a sucrase used industrially for the hydrolysis of sucrose to so-called invert sugar. Lactase is essential for digestive hydrolysis of lactose in milk; many adult humans do not produce lactase and cannot digest the lactose in milk (not a disorder).

The hydrolysis of polysaccharides to soluble sugars is called "saccharification". Malt made from barley is used as a source of β-amylase to break down starch into the disaccharide maltose, which can be used by yeast to produce beer. Other amylase enzymes may convert starch to glucose or to oligosaccharides. Cellulose is first hydrolyzed to cellobiose by cellulase and then cellobiose is further hydrolyzed to glucose by beta-glucosidase. Animals such as cows (ruminants) are able to hydrolyze cellulose into cellobiose and then glucose because of symbiotic bacteria that produce cellulases.

Metal Aqua Ions

Metal ions are Lewis acids, and in aqueous solution they form metal aqua ions of the general formula $M(H_2O)_n^{m+}$. The aqua ions undergo hydrolysis, to a greater or lesser extent. The first hydrolysis step is given generically as

$$M(H_2O)_n^{m+} + H_2O \rightleftharpoons M(H_2O)_{n-1}(OH)^{(m-1)+} + H_3O^+$$

Thus the aqua cations behave as acids in terms of Brønsted-Lowry acid-base theory. This effect is easily explained by considering the inductive effect of the positively charged metal ion, which weakens the O-H bond of an attached water molecule, making the liberation of a proton relatively easy.

The dissociation constant, pK_a, for this reaction is more or less linearly related to the charge-to-size ratio of the metal ion. Ions with low charges, such as Na^+ are very weak acids with almost imperceptible hydrolysis. Large divalent ions such as Ca^{2+}, Zn^{2+}, Sn^{2+} and Pb^{2+} have a pK_a of 6 or more and would not normally be classed as acids, but small divalent ions such as Be^{2+} undergo extensive hydrolysis. Trivalent ions like Al^{3+} and Fe^{3+} are weak acids whose pK_a is comparable to that of acetic acid. Solutions of salts such as $BeCl_2$ or $Al(NO_3)_3$ in water are noticeably acidic; the hydrolysis can be suppressed by adding an acid such as nitric acid, making the solution more acidic.

Hydrolysis may proceed beyond the first step, often with the formation of polynuclear species via the process of olation. Some "exotic" species such as $Sn_3(OH)_4^{2+}$ are well characterized. Hydrolysis tends to proceed as pH rises leading, in many cases, to the precipitation of a hydroxide such as $Al(OH)_3$ or $AlO(OH)$. These substances, major constituents of bauxite, are known as laterites and are formed by leaching from rocks of most of the ions other than aluminium and iron and subsequent hydrolysis of the remaining aluminium and iron.

Alkylation

Alkylation is the transfer of an alkyl group from one molecule to another. The alkyl group may be transferred as an alkyl carbocation, a free radical, a carbanion or a carbene (or their equivalents). An alkyl group is a piece of a molecule with the general formula C_nH_{2n+1}, where n is the integer depicting the number of carbons linked together. For example, a methyl group (n = 1, CH_3) is a fragment of a methane molecule (CH_4). Alkylating agents utilize selective alkylation by adding the desired aliphatic carbon chain to the previously chosen starting molecule. This is one of many known chemical syntheses. Alkyl groups can also be removed in a process known as dealkylation.

In oil refining contexts, alkylation refers to a particular alkylation of isobutane with olefins. For upgrading of petroleum, alkylation produces synthetic C_7–C_8 alkylate, which is a premium blending stock for gasoline.

In medicine, alkylation of DNA is used in chemotherapy to damage the DNA of cancer cells. Alkylation is accomplished with the class of drugs called alkylating antineoplastic agents.

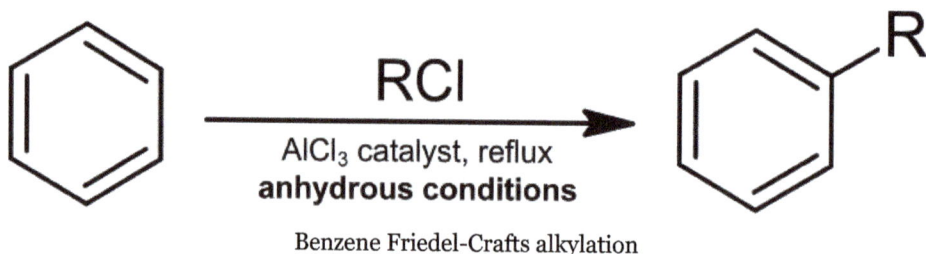

Benzene Friedel-Crafts alkylation

Alkylating Agents

Alkylating agents are classified according to their nucleophilic or electrophilic character.

Nucleophilic Alkylating Agents

Nucleophilic alkylating agents deliver the equivalent of an alkyl anion (carbanion). Examples include the use of organometallic compounds such as Grignard (organomagnesium), organolithium, organocopper, and organosodium reagents. These compounds typically can add to an electron-deficient carbon atom such as at a carbonyl group. Nucleophilic alkylating agents can also displace halide substituents on a carbon atom. In the presence of catalysts, they also alkylate alkyl and aryl halides, as exemplified by Suzuki couplings.

Electrophilic Alkylating Agents

Electrophilic alkylating agents deliver the equivalent of an alkyl cation. Examples include the use of alkyl halides with a Lewis acid catalyst to alkylate aromatic substrates in Friedel-Crafts reactions. Alkyl halides can also react directly with amines to form C-N bonds; the same holds true for other nucleophiles such as alcohols, carboxylic acids, thiols, etc. Trimethyloxonium tetrafluoroborate and triethyloxonium tetrafluoroborate are particularly strong electrophiles due to their overt positive charge and an inert leaving group (dimethyl or diethyl ether).

Electrophilic, soluble alkylating agents are often very toxic, due to their ability to alkylate DNA. They should be handled with proper PPE. This mechanism of toxicity is also responsible for the ability of some alkylating agents to perform as anti-cancer drugs in the form of alkylating antineoplastic agents, and also as chemical weapons such as mustard gas. Alkylated DNA either does not coil or uncoil properly, or cannot be processed by information-decoding enzymes. This results in cytotoxicity with the effects of inhibition the growth of the cell, initiation of programmed cell death or apoptosis. However, mutations are also triggered, including carcinogenic mutations, explaining the higher incidence of cancer after exposure.

Alcohols and phenols can be alkylated to give alkyl ethers:

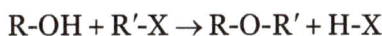

$$R\text{-}OH + R'\text{-}X \rightarrow R\text{-}O\text{-}R' + H\text{-}X$$

The produced acid HX is neutralized with a base, or, alternatively, the alcohol is deprotonated first to give an alkoxide or phenoxide. For example, dimethyl sulfate alkylates the sodium salt of phenol to give anisole, the methyl ether of phenol. The dimethyl sulfate is dealkylated to sodium methylsulfate.

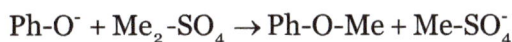

$$Ph\text{-}O^- + Me_2\text{-}SO_4 \rightarrow Ph\text{-}O\text{-}Me + Me\text{-}SO_4^-$$

On the contrary, the alkylation of amines introduces the problem that the alkylation of an amine makes it *more* nucleophilic. Thus, when an electrophilic alkylating agent is introduced to a primary amine, it will preferentially alkylate all the way to a quaternary ammonium cation.

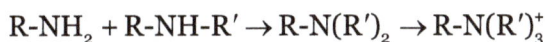

$$R\text{-}NH_2 + R\text{-}NH\text{-}R' \rightarrow R\text{-}N(R')_2 \rightarrow R\text{-}N(R')_3^+$$

If the quaternary ammonium is not the desired product, more circuitous routes such as reductive amination are necessary.

Carbene Alkylating Agents

Carbenes are extremely reactive and are known to attack even unactivated C-H bonds. Carbenes can be generated by elimination of a diazo group. Unlike electrophilic or nucleophilic alkylating agents, carbenes are neutral, and they insert into bonds rather than discard leaving groups. A metal can form a carbene equivalent called a transition metal carbene complex.

Catalysts

Silicotungstic acid is used to manufacture ethyl acetate by the alkylation of acetic acid by ethylene:

$$C_2H_4 + \S CH_3CO_2H \rightarrow CH_3CO_2C_2H_5$$

It has also been commercialized for the oxidation of ethylene to acetic acid:

$$C_2H_4 + O_2 \rightarrow CH_3CO_2H$$

In Biology

Methylation is the most common type of alkylation, being associated with the transfer of a methyl group. Methylation is distinct from alkylation in that it is specifically the transfer of one carbon, whereas alkylation can refer to the transfer of long chain carbon groups. Methylation in nature is typically effected by vitamin B12-derived enzymes, where the methyl group is carried by cobalt. In methanogenesis, coenzyme M is methylated by tetrahydromethanopterin.

Electrophilic compounds may alkylate different nucleophiles in the body. The toxicity, carcinogenity, and paradoxically, cancer cell-killing abilities of different DNA alkylating agents are an example.

Demethylation is the reverse of methylation.

Oil Refining

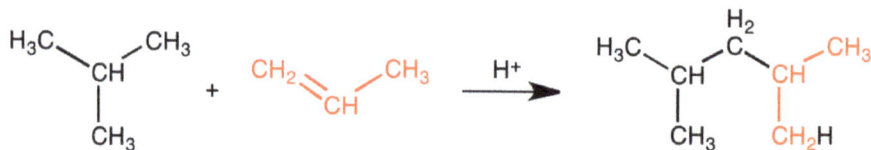

Alkylation of alkenes (shown in red is propene) by isobutane is a major process in refineries to produce higher octane "alkylate" for gasoline blending, in this example yielding isoheptane. It is catalysed by strong acids such as hydrofluoric acid (HF) and sulfuric acid (H_2SO_4)

In a standard oil refinery process, isobutane is alkylated with low-molecular-weight alkenes (primarily a mixture of propene and butene) in the presence of a Bronsted acid catalyst, either sulfuric acid or hydrofluoric acid. In an oil refinery it is referred to as a sulfuric acid alkylation unit (SAAU) or a hydrofluoric alkylation unit, (HFAU). Refinery workers may simply refer to it as the alky or alky unit. The catalyst protonates the alkenes (propene, butene) to produce reactive carbocations, which alkylate isobutane. The reaction is carried out at mild temperatures (0 and 30°C) in a two-phase reaction. Because the reaction is exothermic, cooling is needed: SAAU plants require lower temperatures so the cooling medium needs to be chilled, for HFAU normal refinery cooling water will suffice. It is important to keep a high ratio of isobutane to alkene at the point of reaction to prevent side reactions which produces a lower octane product, so the plants have a high recycle of isobutane back to feed. The phases separate spontaneously, so the acid phase is vigorously mixed with the hydrocarbon phase to create sufficient contact surface.

The product is called alkylate and is composed of a mixture of high-octane, branched-chain paraffinic hydrocarbons (mostly isoheptane and isooctane). Alkylate is a premium gasoline blending stock because it has exceptional antiknock properties and is clean burning. Alkylate is also a key component of avgas. The octane number of the alkylate depends mainly upon the kind of alkenes used and upon operating conditions. For example, isooctane results from combining butylene with isobutane and has an octane rating of 100 by definition. There are other products in the alkylate, so the octane rating will vary accordingly.

Since crude oil generally contains only 10 to 40 percent of hydrocarbon constituents in the gasoline range, refineries use a fluid catalytic cracking process to convert high molecular weight hydrocarbons into smaller and more volatile compounds, which are then converted into liquid gasoline-size hydrocarbons. Alkylation processes transform low molecular-weight alkenes and iso-paraffin molecules into larger iso-paraffins with a high octane number.

Combining cracking, polymerization, and alkylation can result in a gasoline yield representing 70 percent of the starting crude oil. More advanced processes, such as cyclicization of paraffins and dehydrogenation of naphthenes forming aromatic hydrocarbons in a catalytic reformer, have also been developed to increase the octane rating of gasoline. Modern refinery operation can be shifted to produce almost any fuel type with specified performance criteria from a single crude feedstock.

Refineries examine whether it makes sense economically to install alkylation units. Alkylation units are complex, with substantial economy of scale. In addition to a suitable quantity of feedstock, the price spread between the value of alkylate product and alternate feedstock disposition value must be large enough to justify the installation. Alternative outlets for refinery alklylation feedstocks include sales as LPG, blending of C4 streams directly into gasoline to lower the flash point of the product and feedstocks for chemical plants. Local market conditions vary widely between plants. Variation in the RVP (Reid vapor pressure) specification for gasoline between countries and between seasons dramatically impacts the amount of butane streams that can be blended directly into gasoline. The transportation of specific types of LPG streams can be expensive so local disparities in economic conditions are often not fully mitigated by cross market movements of alkylation feedstocks.

The availability of a suitable catalyst is also an important factor in deciding whether to build an alkylation plant. If sulfuric acid is used, significant volumes are needed. Access to a suitable plant is required for the supply of fresh acid and the disposition of spent acid. If a sulfuric acid plant must be constructed specifically to support an alkylation unit, such construction will have a significant impact on both the initial requirements for capital and ongoing costs of operation. Alternatively it is possible to install a WSA Process unit to regenerate the spent acid. No drying of the gas takes place. This means that there will be no loss of acid, no acidic waste material and no heat is lost in process gas reheating. The selective condensation in the WSA condenser ensures that the regenerated fresh acid will be 98% w/w even with the humid process gas. It is possible to combine spent acid regeneration with disposal of hydrogen sulfide by using the hydrogen sulfide as internal fuel in the refinery or elsewhere.

The second main catalyst option is hydrofluoric acid. In typical alkylation plants, rates of consumption for acid are much lower than for sulfuric acid. These plants also produce alkylate with better octane rating than do sulfuric plants. However, due to its hazardous nature, HF acid is produced at very few locations and transportation must be managed rigorously.

Unit Operations in Chemical Industries

Unit operations are very important in chemical industries for separation of various products formed during the reaction. Table give the details of unit operation in chemical process industries.

Table : Unit Operations in Chemical Process Industries

Absorption and stripping	Membrane Process: Reverse osmosis, Ultrafiltration, Dialysis, Electrodialysis, Perevaporation
Adsorption and desorption Pressure Swing adsorption Chromatography	Crushing Grinding, Pulverizing and Screening
Distillation: Batch distillation Flash distillation, Azeotropic distillation, Extractive distillation Reactive distillation	Solid liquid extraction
Evaporation	Striping
Fluidisation	Sublimation
Crystallisation	Solvent extraction
Liquid- Liquid extraction	

Distillation

Distillation has been the king of all the separation processes and most widely used separation technology and will continue as an important process for the foreseeable future [Olujie et al., 2003]. Distillation is used in petroleum refining and petrochemical manufacture Distillation is the heart of petroleum refining and all processes require distillation at various stages of operations.

Membrane Processes

Membrane processes have emerged one of the major separation processes during the recent years and finding increasing application in desalination, wastewater treatment and gas separation and product purification. Membrane technology is vital to the process intensification strategy and has continued to advance rapidly with the development of membrane reactors, catalytic membrane reactor, membrane distillation, membrane bioreactors for wide and varied application [Sridhar, 2009]

Membrane process classified based on driving force. Various type of membrane process and driving force are given in Table.

Table : Membrane Processes

Membrane process	Driving force
Reverse osmosis Ultrafiltration Microfiltration Nanofiltration Dialysis Pervaporation Liquid membrane Electrodialysis Gas Permeation Thermo-osmosis	Pressure difference Pressure difference Pressure difference Pressure difference Concentration difference Concentration difference Concentration difference Electrical potential Concentration difference Temperature difference

Based on lower operating costs, comparable capital cost and only slightly product loss (including fuel), membranes have demonstrated a flexible, cost, effective alternative to amine treating for

some natural gas processing applications [Cook & Losin, 1995]. Gas membrane and its application areas are mention in Table.

Membrane distillation is a membrane separation process, which can overcome the limitation of more traditional membrane process. Membrane distillation has significant advantage over other processes, including low sensitivity to feed concentration and the ability to operate at low temperature [Patli and Patil, 2012]. Various type of membrane processes are mention in Table.

Table : Gas Membrane Application Areas

Common Gas Separation	Application
O_2/N_2	Generation oxygen enrichment, inert gas
H_2 /hydrocarbons	refinery hydrogen recovery
H_2/CO	Syn. gas adjustment
H_2/N_2	Ammonia purge gas
CO_2/hydrocarbons	Acid gas removal from natural gas
H_2O/hydrocarbons	Natural gas dehydration
H_2S/hydrocarbons	Sour gas treating
He/hydrocarbons	Helium separation
He/N_2	HELIUM RECOVERY
Hydrocarbon/ air	Hydrocarbon recovery
H_2O/AIR	Air dehumidification

Table: Various Types of Membrane Processes

Separation Process	Separation Mechanisms	Feed Stream
Microfiltration	Sieving	Liquid or Gas
Ultra-filtration	Sieving	Liquid
Dialysis	Sieving And Sorption Diffusion	Liquid
Reverse Osmosis	Sorption- Diffusion	Liquid
Evaporation	Sorption- Diffusion	Liquid
Gas And Vapour Permeation	Sorption- Diffusion	Liquid or Vapour

Absorption

Absorption is the one of the most commonly used separation techniques for the gas cleaning purpose for removal of various gases like H_2S, CO_2, SO_2 and ammonia. Cleaning of solute gases is achieved by transferring into a liquid solvent by contacting the gas stream with liquids that offers specific or selectivity for the gases to be recovered. Unit operation and is mass transfer phenomena where the solute of a gas is removed from being placed in contact with a nonvolatile liquid solvent that removes the components from the gas.

Solvent: Liquid applied to remove the solute from a gas stream.

Solute: Components to be removed from entering streams.

Some of the Commonly used Solvents are:

Chemical Absorption

Amine Processes: Mono-ethanol amine (MEA), di-ethanol amine (DEA), tri-ethanol amine (TEA), diglycol amine (DGA), methyl diethanol amine (MDEA)

Carbonate Process: K_2CO_3, K_2CO_3+MEA, K_2CO_3 +DEA, K_2CO_3+arsenic trioxide

Physical Absorption

Polyethylene Glycol Dimethyl Ether (Selexol), N-methyl pyrrolidine,NMP (Purisol), Methanol (Rectisol), Sulphonane mixed with an alkanolamine and water (sulfinol).

Adsorption

Adsorption technology is now used very effectively in the separation and purification of many gas and liquid mixtures in chemical, petrochemical, biochemical and environmental industries and is often a much cheaper and easier option than distillation, absorption or extraction. Some of the major applications of adsorption are gas bulk separation, gas purifications, liquid bulk separation, liquid purifications [Keller II, 1995]. One of the most effective method for recovering and controlling emissions of volatile organic compounds is adsorption Some of the commercial adsorbent s are silica gel, activated carbon, carbon molecular sieve, charcoal, zeolites molecular sieves, polymer and resins, clays, biosorbents. some of the key properties of adsorbents are capacity, selectivity, regenerability, kinetics, compatibility and cost [Knaebel, 1995]. Some of the methods used for regeneration of adsorbent are thermal swing, pressure swing, vacuum (special case of pressure swing), purge and gas stripping, steam stripping [Crittenden, 1988]. Commercial adsorption processes is given in Table. Some of the important criteria of good adsorbent are [Keller II, 1995].

(1) it must selectivity concentrate one or more components called adsorbate to from their fluid phase levels

the ability to release adsorbate so that adsorbent can be reused,

as high as possible delta loading the change of weight of adsorbate per unit weight of adsorbent between adsorbing and desorbing steps over a reasonable range of pressure and temperature

Table : Commercial Adsorption Processes

Sorbex process	Application
Parex	Separation of paraxylene from mixed C_8 aromatics isomers
MX sorbex	Meta xylene from mixed C_8 aromatics isomers
Molex	Linear paraffins from branched and cyclic hydrocarbons
Olex	Olefins from paraffins
Crsex	Para cresol or meta cresol isomers
Cymex	Para cymene or meta cymene from cymene isomers
Sarex	Fructose from mixed sugar
UOP ISOSIV processor	separation of normal paraffins from hydrocarbon mixture
Kerosene Isoiv process	For separation of straight chain normal paraffins from the kerosene range(C10-C18) used for detergent industry

Pressure swing adsorption (PSA) is based on the principle of relative adsorption strength, is a milestone in the science of gas separation [Shiv Kiran and Chakravarty, 2002]. Some of the commercial application of PSA are air drying, hydrogen purification, bulk separation of paraffins, air separation for oxygen and nitrogen production,

Chromatography is a sorptive separation process. in choromatography feed is introduce in column containing a selective adororbent(stationary phase) and separated over the length of the column by the action of a carrier fluid (mobile phase)that is continually supplied to the column following the introduction of the feed. The separation occurs as a result of the different partitioning of the feed solutes between the stationary phase. The separated solutes are recovered at different time in the effluent from the column [Rangrajan,2010].

Crystallization Process

Crystallization processes are used in the petroleum industry for separation of wax. The process involves nucleation, growth, and agglomeration and gelling. Some of the applications of crystallization is in the separation of wax, separation of p-xylene from xylenes stream. Typical process of separation of p-xylene involves cooling the mixed xylene feed stock to a slightly higher than that of eutectic followed by separation of crystal by centrifugation or filtration.

Liquid –Liquid Extraction

Liquid –liquid extraction has been commonly used in petroleum and petrochemical industry for separation of close boiling hydrocarbons. Some of the major applications are:

- Removal of sulphur compound from liquid hydrocarbons

- Recovery of aromatics from liquid hydrocarbon

- Separation of butadiene from C_4 hydrocarbons

- Extraction of caprolactam

- Separation of homogenous aqueous azeotropes

- Extraction of acetic acid

- Removal of phenolic compounds from waste water

- Manufacture of rare earths

- Separation of asphaltic compounds from oil

- Recovery of copper from leach liquor

- Extraction of glycerides from vegetable oil

Some of the important property of a good solvent

- High solvent power/capacity

- High selectivity for desired component

- Sufficient difference in boiling points of the solvent and the feed for effective separation

- Low latent heat of evaporation an specific heat to reduce utility requirement

- high thermal an chemical stability

- Low melting point

- Relatively inexpensive

- Non toxic and non –corrosive

- Low viscosity low interfacial tension

Technological Development in Unit Operations-

- Distillation, Azeotropic, extractive distillation, reactive distillation, membrane distillation

- Random packing to Structured Packing

- Single and two pass to Multiple down comer

- Rasching rings and berl saddles to Intalox sadles, pall rings, nutter rings, half rings, super rings,Fleximax

- Pan park to Wire gauge packing, Goodloe, Mellpark, Flexipack, Gempack, Intalox

- Fixed bed to Fluidised bed reactor

- Conventional reactor to Micro reactor

- Ball mill grinding to Vertical roller mill and press roll Mill

- Open circuit grinding to Closed circuit grinding

- Batch digester to continuous digester

- Low speed and low capacity cipper to High speed chipper and high capacity chipper

- Low speed Paper machine to high speed machine

- Drum displacer, Pressure diffuser, Displacement presses, Combined deknotting and Fine screening,

- High temperature screening before washing, Reverse cleaners

- Adsorption(Olex, Parex and Molex), Crystallisation and Membrane separation processes

- Solvent extraction processes and New solvents

- Conventional distillation Short path distillation, divided wall column

- Conventional bubble cap, sieve plate to valve tray

- Random packing to structured packing

- Axial flow reactor to radial flow reactor

- Conventional instrumentation to smart (intelligent) instrumentation

Organic Chemical Industries

Chemical process industry play an important role in the development of a country by providing a wide variety of products, which are being used in providing basic needs of rising population which is 6.4 billion globally and 1.2 billion in India in 2012. Chemical process industries uses raw material derived from petroleum and natural gas, salt, oil and fats, biomass and energy from coal, natural gas and a small percentage from renewable energy resources. Although initially manufacture of organic chemicals initially started with coal and alcohol from fermentation industry, however later due to availability of petroleum and natural gas dominated the scene and now more than 90% of organic chemicals are produced from petroleum and natural gas routes. However, rising cost of petroleum and natural gas and continuous decrease in the reserves has spurred the chemical industry for alternative feedstock like coal, biomass, coal bed methane, shale gas, sand oil as an alternate source of fuel and chemical feedstock.

Energy Resources

Energy play vital role an important role for the development of any country and to meet the challenges due to increasing population it has become one of the very important to optimize its use and look for alternative energy resources. Coal remains the dominant source of energy meeting 52.4% of India's prime energy needs while oil and natural gas met 41.6% of energy requirement in 2008-09. Power sector accounted for 77% of the non-coking coal off-take. As per planning commission projections till 2032, coal will continue to have a dominant share meeting over 50% of primary commercial requirement .

World and India Energy consumption scenario is given in Figure [Hindu Industrial directory, 2007]. The world energy consumption had projected to increase by 58% over a 24 year period from 2001 to 2025. The total energy use projected to grow from 404 in 2001 to 640

quadrillion BTU in 2025 [Energy outlook, 2003]. India's requirement for fossil fuels by 2030 is estimated by various agencies is in the range of 337 to 462 million tones of oil, 99 to 184 million tones oil equivalent of gas and 602 to 954 million tones of coal Indian energy's .

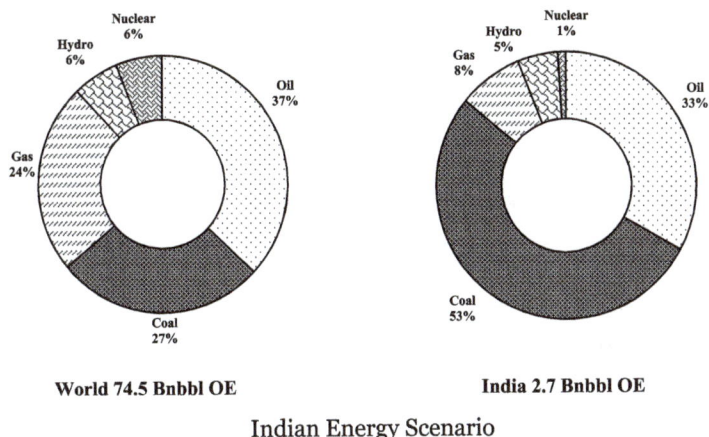

World 74.5 Bnbbl OE   India 2.7 Bnbbl OE

Indian Energy Scenario

Raw Materials for Organic Chemical Industries

Although initially manufacture of organic chemicals initially started with coal and alcohol from fermentation industry, however later due to availability of petroleum and natural gas dominated the scene and now more than 90% of organic chemicals are produce from petroleum and natural gas routes. However, rising cost of petroleum and natural gas and continuous decrease in the reserves as spurred the chemical industry for alternative feedstock like coal, biomass, coal bed methane, shale gas, sand oil as alternate source of fuel and chemical feedstock. Table gives the details of raw materials for chemical process industries. Raw materials for chemical industries are classified as primary raw materials and basic intermediates. Although major organic chemicals are produced from petroleum feed stock, however alternative raw materials are available which are getting attention. Detail of feedstock for organic chemical industries is shown in Figure. Table shows the details of natural gas and petroleum fractions as petrochemicals feedstock. Alternative Routes to Principal Organic Chemicals is given Table.

Table : Raw Material for Chemical Process Industries

Primary Raw Materials:

Gaseous	Natural gas, condensate, refinery gases, coal Bed methane, gas hydrate
Liquids	Naphtha, kerosene, gas oil, middle distillates
Solids	Coal, coke, wax, residues
Oils and fats	Tallow and coconut oil, palm oil and other oil
Biomass	Alcohol, paper, energy,
Salt	Chlorine, caustic soda, soda ash
Sulphur	Sulphuric acid, fertilizer,
Lime stone	Cement, lime

Basic Intermediates

Paraffins	Methane , propane, butane and higher hydrocarbons
	Ethylene, propylene, butadiene , alcohol, vinyl chloride
Olefins and derivatives	Ethylene, propylene,, butadiene , alcohol, vinyl chloride
Aromatics	Benzene Toluene Ethyl benzene, Xylenes, Naphthalene

Secondary Intermediates

Monomer: Caprolactam, adipic acid, hexamethylene diamine, terephthalic acid and acrylonitrile for synthetic fibres, intermediates for dye stuff industry and pesticides.

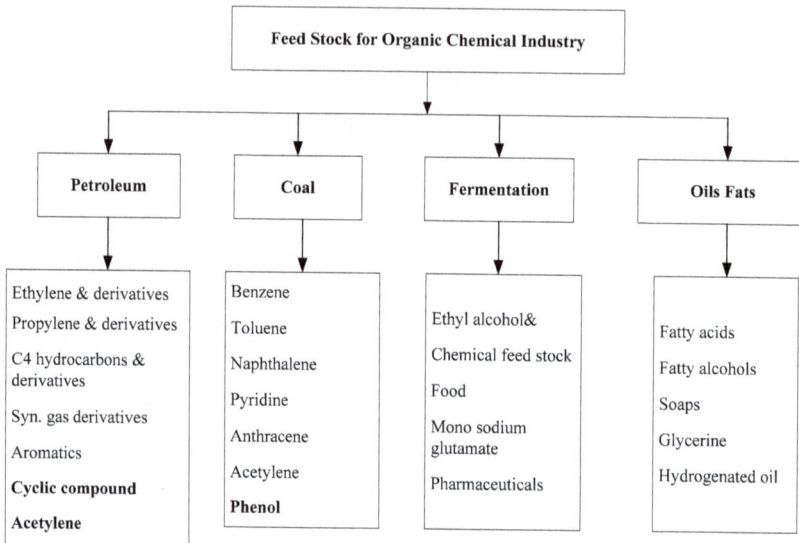

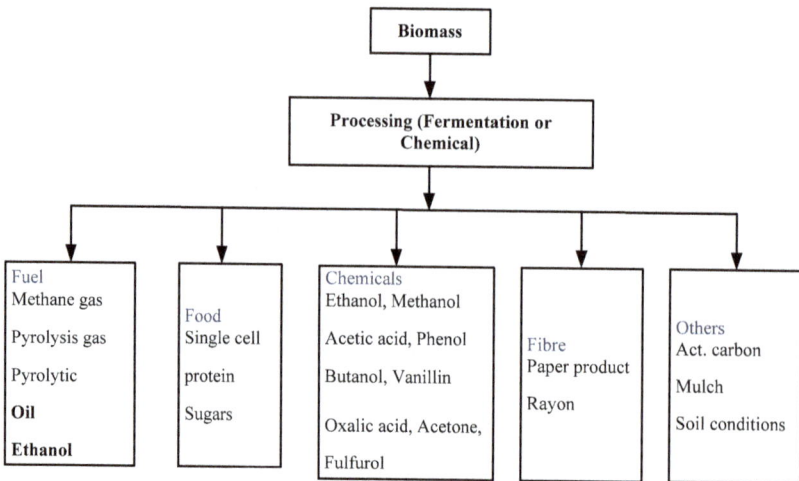

Feed Stock for Organic Chemical Industry

Petroleum | Coal | Fermentation | Oils Fats

Ethylene & derivatives
Propylene & derivatives
C4 hydrocarbons & derivatives
Syn. gas derivatives
Aromatics
Cyclic compound
Acetylene

Benzene
Toluene
Naphthalene
Pyridine
Anthracene
Acetylene
Phenol

Ethyl alcohol&
Chemical feed stock
Food
Mono sodium glutamate
Pharmaceuticals

Fatty acids
Fatty alcohols
Soaps
Glycerine
Hydrogenated oil

Biomass

Processing (Fermentation or Chemical)

Fuel
Methane gas
Pyrolysis gas
Pyrolytic
Oil
Ethanol

Food
Single cell
protein
Sugars

Chemicals
Ethanol, Methanol
Acetic acid, Phenol
Butanol, Vanillin
Oxalic acid, Acetone,
Fulfurol

Fibre
Paper product
Rayon

Others
Act. carbon
Mulch
Soil conditions

Feed Stock for Organic Chemical Industries

Table : Natural Gas and Petroleum Fractions as Petrochemicals Feedstock

Petro-leum Frac-tions and Natural Gases	Source	Composition	Intermediate Processes	Intermediate Feedstock
Refinery Gases	Distillation, catalytic cracking, catalytic reforming	Methane, ethane, propane, butane, BP upto $25^{\circ}C$	Liquefaction, cracking	LPG, ethylene propylene, butane, butadiene.
Naphtha	Distillation and thermal & catalytic cracking, visbreaking	C_4-C_{12} hydrocarbon, BP 70 - $200^{\circ}C$	Cracking, reforming, alkylation, disproportiona- tion, isomerisation	Ethylene, propylene, butane, butadiene, benzene, toluene, xylene
Kero-sene	Distillation and secondary conversion processes	C_9-C_{10} hydrocarbon, BP 175-$275^{\circ}C$	Fractionation to obtain $C_{10}-C_{14}$ range hydrocarbon	Linear n C_{10} - n C_{14} alkanes
Gas Oil	Distillation of crude oil and cracking	$C_{10}-C_{25}$ hydrocarbons BP $200-400^{\circ}C$	Cracking	Ethylene, propylene, butadiene, butylenes
Wax	Dewaxing of lubricating oil	C_8-C_{56} hydrocarbon	Cracking	C_6-C_{20} alkanes
Pyrolysis Gasoline	Ethylene cracker	Aromatic, alkenes, dienes, alkanes, cy-cloalkane	Hydrogenation distillation, extraction, crystallisation, adsorption	Aromatics
Natural Gases & Natural Gas Condensate	Gas fields and crude oil stabilisation	Hydrogen, methane, ethane, propane, pentane, aromatics	Cracking, reforming, separation	Ethylene, propylene, LPG, aromatics, etc.
Petro-leum Coke	Crude oil	Carbon	Residue upgradation processes,gasifi- cation	Carbon electrode, acetylene, fuel

Table: Alternative Routes to Principal Organic Chemicals

Chemicals	Petroleum Source	Alternate Source
Methane	Natural gas, Refinery light gases (de-methaniser overheads)	Coal, as by-product of separation of coke oven gases (1920-30) or of coal hydrogenation (1930-40)

Ammonia	Methane Light liquid hydrocarbons	From coal via water gas (1910-20)
Methyl alcohol	Methane Light liquid hydrocarbons	From coal via water-gas (1920-30); from methane (from coal) by methane-stream and methane oxygen processes (1930-40)
Ethylene	Pyrolysis of gaseous liquid hydro-carbons	Dehydration of ethyl alcohol (original route). By-product in fractional distillation of coke oven gas (1925-35). Hydrogenation of acety-lene (1940-45)
Acetylene	Ethane	Calcium carbide (original process). methane from coal by partial combustion and by arc process (1935-45)
Ethylene glycol	Ethylene	From ethylene made as above (1925). In America, from coal via carbon-monoxide and formaldehyde (1935-40)
Ethyl alcohol	Synthetic ethyl alcohol ,	Fermentation of molasses (original route)
Acetaldehyde	Co-product of paraffin gas oxida-tion. Direct oxidation of ethylene	Fermentation of ethyl alcohol, or acetylene from carbide (1900-10)
Acetone	Propylene	Wood distillation (original process). Pyroly-sis of acetic acid (1920-30) or by acetylene-stream reaction (1930-40)
Glycerol	Propylene	By-product of soap manufacture (original process)
Butadiene	2-Butenes Butane Synthetic ethyl alcohol By-product of ethylene by pyrolysis of liquid hydrocarbons	Ethyl alcohol (1915); acetaldehyde via 1:3- butanediol (1920-30); acetylene and formaldehyde from coal via 1:4-butanediol (1940-45); from 2:3-Butanediol by fermen-tation (1940-45)
Aromatic hydro-carbons	Aromatic-rich and naphthenic-rich fractions by catalytic reform-ing and direct extraction or by hydro-alkylation	By-products of coal-tar distillation

ROUTES TO PRODUCE CHEMICALS

- Steam Reforming and Partial Oxidation (Synthesis gas ($CO + H_2$ and H_2 $\& N$) to produce synthesis gas

- Cracking and Pyrolysis to olefins (C_2H_4, C_3H_6, C_4H_8 and olefins)

- By-products (Pyrolysis gasoline and Higher liquids, Gas condensate) for aro-matics

- Catalytic Reforming to produce mainly BTX from naphtha.

- Dehydrogenation of Paraffin(Paraffin: ethane, paraffin) to produce olefin

- Petrocoke and Biomass gasification

- GTL(Gas to liquid), MTO (Methanol to Olefin),

- Coal to liquid fueland coal to chemicals

- Dehydrogenation (olefin) and alkylation (alkylate) from kerosene for LAB

- Saponification of oil and fats and recovery of chemical from glycerin

NATURAL GAS AS CHEMICAL FEED STOCK

- Chemicals from methane

- Chemicals from $C_2 - C_4$

- $C_5 +$ (natural gasoline)

- Methane/total natural gas

ROUTES FOR NATURAL GASES AS CHEMICAL FEED STOCK

- Cracking of natural gas to olefins, C_4 and C_5 chemicals

- Steam reforming and Partial oxidation for synthesis gas

- Conventional steam reforming

- Partial oxidation (POX)

- Catalytic partial oxidation (CPO)

- Combined reforming

- Combined reforming with performer

- Gas-heating reforming (GHR)

- Auto-thermal reforming

- Combined auto-thermal reforming (CAR)

- Kellogg heat reforming exchanger system (KRES)

  ❖ Cyclar process: For production of aromatics fro natural gas (Propane and butane)

  ❖ Oxidative coupling of methane (natu al gas) to olefins

Naphtha as Chemical Feed Stock

Naphtha is the most versatile chemical feedstock and its use depends on composition, boiling range, end use market requirements. Naphtha remains prominent feedstock

(52%) for olefin production from steam cracker. Feedstock of olefins is shown in Figure. Catalytic reforming of naphtha produces aromatics which is important chemical feed stock for organic chemical industries for producing synthetic fibre, pesticides explosive, dyes intermediate, plasticizer, solvent etc. Some of the routes for conversion of naphtha to petrochemicals are:

Steam reforming/ Partial oxidation of naphtha: For production of synthesis gas and derivatives

Cracking of naphtha: For production of olefins, C4 and C5 hydrocarbons, pyrolysis gasoline for aromatic production

Catalytic reforming of naphtha: For production of aromatics- benzene, toluene, xylenes, Kerosene as Feed Stock for Lab

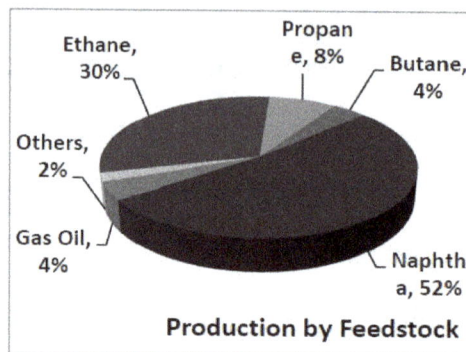

Olefin Feed Stock

n-Paraffins from SR Kerosene : n-Paraffins are extracted using adsorptive separation by molecular sieves. These paraffins are excellent feedstock for LAB. Various steps involved are: Kerosene Pre-fractionation: To tailor the kerosene to desired carbon range

Hydrotreatment: To remove sulfur, nitrogen and olefins and oxygenates which might poison the Molex adsorbent.

Alternate Feed Stock for Chemical Industry

In view of dwindling fossil fuel sources and increasing cost of crude and volatile market oil, there is tremendous activity all over world to utilize alternative feedstock's alternative feedstock includes biomass and algae, coal, petrocoke, waste plastic for production of synthesis gas, olefin, methanol, ethanol and derivatives, naphtha.

Biomass

Biomass resources like crop residues, forage, grass, crops, wood residues, forest residues, short rotation energy crops and cellulosic components of municipal solid waste can be use as alternative feedstock for production of synthesis gas, ethanol, and naphtha through FT process. Alternative energy resources will play a growing role and bio-fuels mainly ethanol are expected to grow rapidly, reaching about 2% of total liquid supplies by 2030 [Singh et al., 2008]. Some of the routes for conversion of biomass to heat & power, transport fuels, bioethanol is given in Figure [Banerjee et al., 2011].

The constantly depleting resources of conventional energy and the steeply escalating price of fossil fuels have led to the need of alternate energy sources. Second generation production of bioethanol production is gaining increasing impetus due to abundant availability, high cellulose d hemicelluloses content of lingo-cellulose materials [Tuil et al., 2008]. Biotechnological route for bioethanol production utilizing lingo cellulose material involves delignification, sacchrification and fermentation. The most common process of bioethanol production from sacchrified lingo cellulosics involves hydrolysis of cellulose and fermentation in the same reactor [Lo, 2009]. Options for Conversion of Biomass to Fuel and Power and Chemical feed stock is given in Figure.

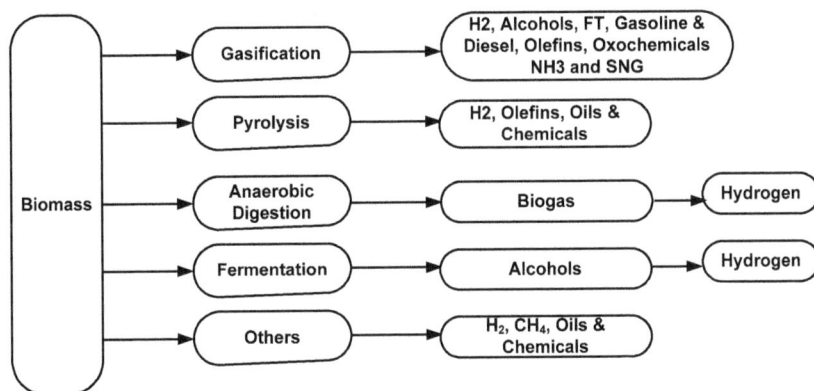

Biomass conversion technologies for chemical production

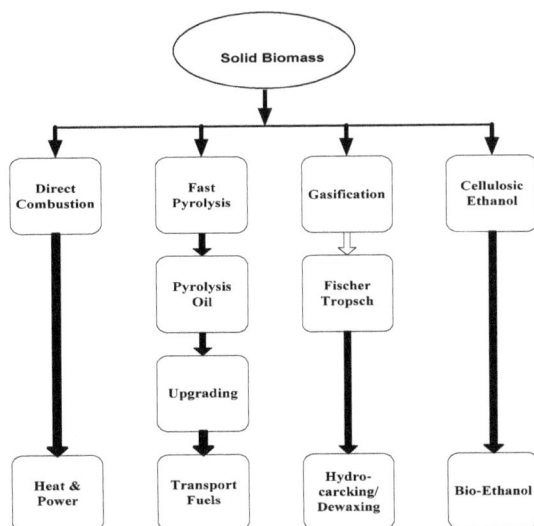

Options for Conversion of Biomass to Fuel and Power and Chemical feed stock

Ethanol

There are three types of feedstocks for ethanol production [Tuli et al. 2008]

Sugars: Molasses, cane sugar, beat sweet sorghum and fruits

Starches: Corn, wheat, rice, potatoes, cassava, sweet potatoes etc Lignocellosic: Straw, bagasse, other agricultural residues, wood, energy crops Algae: Ethanol production

Biomass can be also used a feedstock for methanol production and hydrogen through synthesis gas produced from biomass gasifiers.

Ethanol from Algae

Direct to Ethanol technology, a novel technology developed by Algenol is in Pilot plant testing at Dow Chemical's Freeport, Texas to produce ethanol by photosynthesis from CO_2, H_2O and sunlight instead of producing carbohydrates [Dutta, 2011]

Coal

Coal is another promising feed stock as huge amount of coal reserves is available in India and other part of the world. Based on the production of coal gasification unit it will be possible to to produce large number of chemicals. Possibility Coal as a source of petrochemicals, which again explored all over the world [Handa and Ganesh, 2010]

Coal was the original feed stock for production of large number of chemicals through coke oven plants, synthesis gas from gasification, acetylene fro calcium carbide route. However, due to availability of petroleum based raw material presently more than 90% of chemicals are produce from petroleum and natural gas. Due to volatile market of crude oil and dwindling petroleum resources, coal is emerging as alternative feedstock for chemical industry as huge coal reserves are available all over the world. Various routes for utilization of coal as chemical feedstock and fuel are [Duchesne, 2011, Patil, 2009, Furimsky, 1999]:

- Gasification

- Coal to fuel through FT process

- Coal to methanol technology, Liquid phase Methanol process from coal (LP-MEOHTM)

- Methane to MTO plus Olefin cracking process (OCP)

- Coal to Olefin technology

- Coal to Plastic technology

Petrocoke

Due to the use of heavy crude oil, huge amount of petrocoke is being produce from the thermal cracking process in the refinery. Although petrocoke is being used as fuel in cement industry however it can be a promising raw material for production of synthesis gas, hydrogen, methanol through petrocoke gasification. Through FT synthesis the synthesis gas can be converted to fuel also. Utilization of petrocoke offers an alternative to handle high sulfur and metal containing residues in a refinery with value addition. Reliance is already in process of implementing petrocoke gasification to utilize its petrocoke.

References

- Richens, D. T. (1997). The chemistry of aqua ions: synthesis, structure, and reactivity: a tour through the periodic table of the elements. Wiley. ISBN 0-471-97058-1

- Misono, Makoto (2009). "Recent progress in the practical applications of heteropolyacid and perovskite catalysts: Catalytic technology for the sustainable society". Catalysis Today. 144 (3-4): 285–291. doi:10.1016/j.cattod.2008.10.054

- Greenwood, Norman N.; Earnshaw, Alan (1997). Chemistry of the Elements (2nd ed.). Butterworth-Heinemann. p. 384. ISBN 0-08-037941-9

- Michael Röper, Eugen Gehrer, Thomas Narbeshuber, Wolfgang Siegel "Acylation and Alkylation" in Ullmann's Encyclopedia of Industrial Chemistry, Wiley-VCH, Weinheim, 2000. doi:10.1002/14356007.a01_185

- March Jerry; (1985). Advanced Organic Chemistry reactions, mechanisms and structure (3rd ed.). New York: John Wiley & Sons, inc. ISBN 0-471-85472-7

Coal and its uses

Coal is a necessary component for industrial use. It is used as a fuel owing to its combustible nature. It is processed using various methods such as coal carbonization, hydrogenation of coal, coal gasification, etc. This section discusses the methods of coal processing in a critical manner providing key analysis to the subject matter.

Coal Chemicals

Coal is used as fuel for electric power generation, industrial heating and steam generation, domestic heating, rail roads and for coal processing.

Coal composition is denoted by rank. Rank increases with the carbon content and decreases with increasing oxygen content.

Many of the products made by hydrogenation, oxidation, hydrolysis or fluorination are important for industrial use. Stable, low cost, petroleum and natural gas supplies has arose interest in some of the coal products as upgraded fuels to reduce air pollution as well as to take advantage of greater ease of handling of the liquid or gaseous material and to utilize existing facilities such as pipelines and furnaces.

Coking of Coal

Raw material is Bituminous coal. It appears to have specific internal surface in the range of 30 to 100m²/g.

Generally one ton of bituminous coal produces

- ➢   1400 Ib of coke.

- ➢   10 gallons of tar.

Chemical Reaction

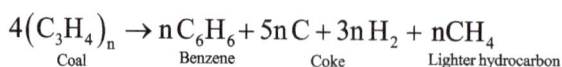

$$4\left(C_3H_4\right)_n \rightarrow n\,C_6H_6 + 5n\,C + 3n\,H_2 + n\,CH_4$$

Coal   Benzene   Coke   Lighter hydrocarbon

Process flow sheet: Illustrated in Figure

Flow sheet of coking of coal

Functional Role of Each Unit

(a) Coal crusher and screening:

- At first Bituminous coal is crushed and screened to a certain size.

- Preheating of coal (at 150-250°C) is done to reduce coking time without loss of coal quality.

- Briquetting increases strength of coke produced and to make non coking or poorly coking coals ...to be used as metallurgical coke.

- Blending prevents damage to oven when high pressure develops.

(b) Coke oven:

- Oven has usually 0.3 to 0.6m width.

- Here coal is kept for 17 hrs, heat is supplied completely from the flues on the sides.

- The oven temperature is usually 1100°C.

- As temperature is increased, the fluidity of mass reaches maximum and begins to solidify to ...form coke.

- Indirect heat transfer is done to prevent burning of coke and formation of carbon dioxide.

- Air is prevented so that no burning takes place inside the oven.

- After carbonization, the oven doors are opened and the red hot coke is pushed to quenching ...car.

- Heat recovery systems and vertical flues are attached to supply process heat to the coking ...chamber.

- The by-product vapors and gases like light-oil, NH3, tar, coke oven gas which are combinedly known as foul main stream are collected and passed through condenser.

(c) Gas generator:

- Coke breeze comes to generator and is heated with air.

- Producer gas produced is sent back to the oven for maintaining high temperatures.

- The air supply for the furnaces is preheated by the hot exit gas.

(d) Quenching car:

- The red hot coke from oven is pushed to quenching car.

- On quenching the fluid coke gets crystallized.

- It is then dumped on a sloping wharf for cooling and drying.

(e) Coke crusher and screening:

- After draining off the water from coke, coke is crushed and screened to be used as a fuel.

- The screened coke is then sent to loading units.

- Coke breeze (or fines) is generated while crushing of coke.

(f) Condenser:

- The foul main is cooled by indirect heat transfer with water for condensation of tar.

- The lighter components are sent to tar extractor for further removal of tar.

(g) Tar extractor:

- The tar along with some lighter components is sent to tar separator.

(h) Reheater:

- The tar free aromatics and NH_3 are heated and sent to saturator.

(i) Tar separator:

- Tar is completely separated and sent to tar storage.

- The lighter components are sent to lime still.

(j) Lime still:

- The lighter components from the tar separator are treated with $Ca(OH)_2$ and steam.

- Phosphates are collected as sludge from the bottom and the aromatics and NH_3 are sent to ...saturator.

k) Saturator:

- The input stream is scrubbed with H_2SO_4.

- The ammonia is separated out from aromatics.

- The aromatics are sent to final cooler.

(l) Centrifuge:

- $NH_3 + H_2SO_4 \rightarrow (NH_4)_2SO_4$

- Ammonium sulfate which is solid and is collected in a centrifuge.

(m) Cooler:

- The aromatics along with sulfur are cooled up for better scrubbing.

(n) Oil scrubber:

- The straw oil is used as scrubber and this will remove aromatics, which are then sent to light oil ..still.

(o) Oil still:

- The straw oil is recycled back to scrubber.

(p) Oxide purifier or wet purifier:

- Aromatic free gas ie. Coal gas from the scrubber comes in this unit.

- The coal gas is made H2S free by oxide purifier or wet purifier.

- The sulfur free gas is collected in gas holder and is used as fuel gas in furnaces.

Coke Oven Plant

Due to the development of iron and steel industry coke oven plant has become an integral part of iron and steel industry. Due to increasing demand of iron and steel, there has been a considerable increase in the coke oven capacity which resulted increase output of coal chemicals.

Two types of coke manufacturing technologies use are:

- Coke making through by product recovery

- Coke making through non-recovery/ heat recovery

In India, building of coke oven batteries was initiated in the beginning of the ninth century, now about 3000 ovens are in operation/ construction in the coke oven plant. By the year 2011-12 ,the world coking coal requirement will be about 433 million metric tones in which India's requirement is estimated to about 54 million tones. By product from coal gasification plant includes coke, coal tar, sulphur, ammonia. Coal tar distillation produces tar, benzol, cresol, phenol, creosote.

Coking Coals: Blast furnace requires coke of uniform size, high mechanical strength, and porosity with minimum volatile matter and minimum ash. Coking coal may be dived I on the basis of their coking properties: prime coking coal, medium coking coal, semi coking coal. The prime coking coal produce strong metallurgical coke while coals of other groups yield hard coke only the concentration of moisture ash, sulphur and sometime phosphorous and ash fusion temperature are important in determining the grade of coking coal since they influence the quality of coke produced. Low moisture, ash, sulphur and phosphorous content in the coal are desirable for production of good quality coke. The desired analysis of typical coal charge to coke oven is.

Ash content	: 16% ±0.5%
Moisture	: 6-7%
Volatile matter	: 22-25%
Fixed carbon	: 58-60%%
Sulphur	: 0.56%
Phosphorous	: 0.09%

Some of the other factors affecting quality of coke are rank of coal, particle size, bulk density, weathering of coal, coking temperature and coking rate, soaking time, quenching practice.

Coke Oven Plant: Various sections in coke oven plant are given in Table.

Table : Various section in coke oven plant are

Coal Handling Plant and Coal Preparation Section	To prepare coal blend suitable for carbonization. various steps involved are unloading and storage of coal, blending of coal of various grade, coal crushing and transport to coal storage tower
Partial briquetting	To prepare briquette of coal to charge along with coal into the coke oven.
Coke oven Batteries	To convert coal into coke by carbonizing coal in absence of air. The process steps involved are coal charging and coal carbonisation

Coke sorting Plant	Crushing and sorting of coke to suitable size for use in blast furnace. The steps involved are coke pushing, coke quenching, coke crushing/ screening
Coke oven gas re-covery	Collection and cleaning of coke oven gas and recovery of by products. . This involves gas cooling, tar recovery, desulphurization of coke oven gas , recovery of ammonia, recovery of light oil
Ammonia recovery and Ammo-niumSulphate Production.	Recovery of ammonia and neutralization with sulphuric acid or nitric acid in case of ammonium nitrate/ calcium ammonium nitrate.
Waste water treat-ment	Treatment of phenolic waste water

Coal Handling Plant and Coal Preparation Section

Coal needs to be stored at various stages of the preparation process, and conveyed around the coal preparation section. Crushing and screening are the important part of coal handling plant. Crushing reduces the overall size of the coal so that it can be more easily processed and handled. Screens are used to ranges the size of coal. Screens can be static, or mechanically vibrated. Dewatering screens are used to remove water from the product.

Partial Briquetting: Briquettes of low grade coal are prepared using a binder (pitch/ pitch+tar) upto 2 to 3.0% of charge. This partial briquette of coal is charged with coal into coke oven which significantly improve the quality of coke.

Coke oven Batteries: Coke oven plant consists of Coke oven batteries containing number of oven (around 65 ovens in each battery). The coal is charged to the coke oven through charging holes. The coal is then carbonized for 17-18 hours, during which volatile matter of coal distills out as coke oven gas and is sent to the recovery section for recovery of valuable chemicals. The ovens are maintained under positive pressure by maintaining high hydraulic main pressure of 7 mm water column in batteries. The coking is complete when the central temperature in the oven is around 950-1000 oC. At this point the oven is isolated from hydraulic mains and after proper venting of residual gases, the doors are opened for coke pushing. At the end of coking period the coke mass has a high volume shrinkage which leads to detachment of mass from the walls ensuring easy pushing. The coke is then quenched and transferred to coke sorting plant.

The control of oven pressure is quite important because lower pressure leads to air entry while higher pressure leads to excessive gassing, leakage of doors, stand pipe etc. Proper leveling of coal is important and care is taken so that free board space above (300 mm) is maintained to avoid choking.

Coke oven plants are integral part of a steel plant to produce coke, which is used as fuel in the blast furnace. Coke oven plant produces important by product coal chemical tar, ammonia, crude benzoyl which is fractionated to produce aromatics-benzene toluene, xylene. Typical flow diagram of coke oven plant is shown in Figure. The Coke Oven Plant consists of following sections:

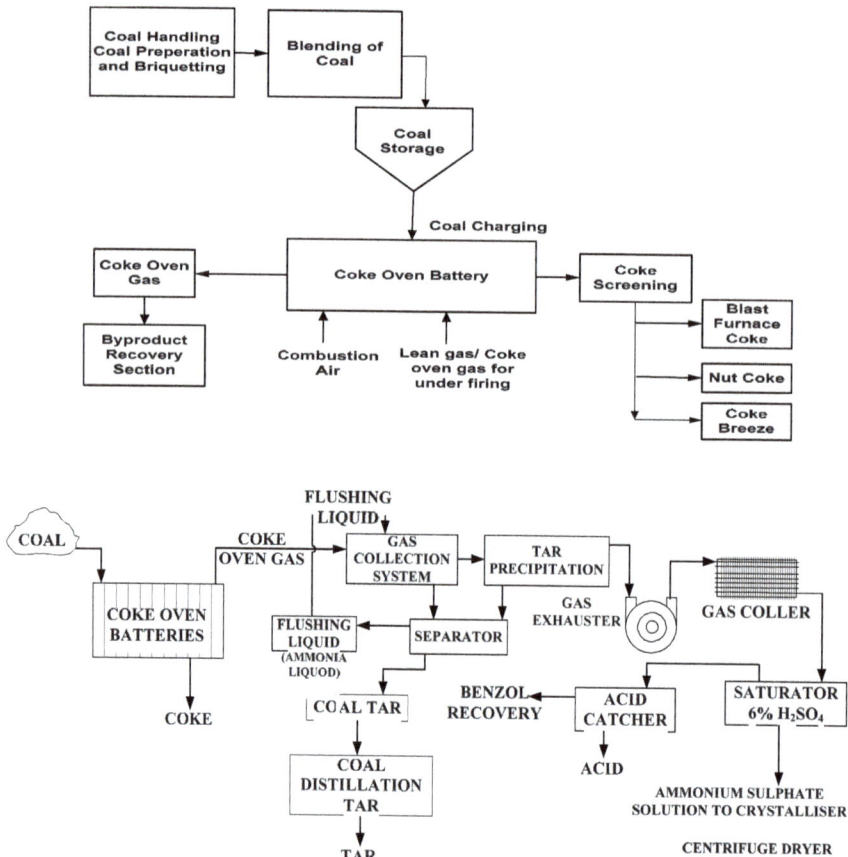

Flow Diagram of Coke Oven Plant

Coke Oven Batteries

Coke oven are used to convert coal into coke by carbonizing coal in absence of air and there by distilling the volatile matter out of coal. Coke is taken as product which is use as fuel and as a reducing agent in smelting iron ore in a blast furnace and coke oven gas as byproduct is treated for recovery of coal chemicals. The coke oven temperature is keep as high as 2000 °C. Crushing and screening of coke is done to obtain suitable size for use in blast furnace. Typical analysis of blast furnace coke is given in Table.

Table : Typical Analysis of Blast Furnace Coke

Sr. No.	Parameter	Value in percentage
1	Moisture	3.5-6%
2	Ash	15.5-17.0%

3	V.M.	<1.00
4	Sulphur	0.65%
5	Fixed carbon	79-81%

Coke Oven Gas plant

Coke oven gas produce during the process of coking of coal are used in coke oven gas plant for the recovery of various valuable chemicals like tar, ammonia and benzoyl. Typical analysis of coke oven gas is mention in Table. These chemicals are recovered and gas is cleaned this comprises following sections:

By-product from Coke Oven Plant

The high temperature carbonization is used for production of coke for use in blast furnace. Various by-products obtained from coal carbonization are crude tar, crude benzoyl, and ammonia. Typical yield of some important byproduct are: Tar 3.2%, ammonium sulphate 1.1%, crude benzoyl 0.9%.

Gas Condensation section: Coke oven gas containing water vapours and chemical products of coking (tar, ammonia, benzoyl etc. at temperature about 750-800°C from the coke oven plant is cooled to temperature of 80-82°C. During gas cooling 65-70% of the tar is condensed. Further cooling of gas, the water vapors and the remaining part of the tar get condensed along with some ammonia and other chemicals.

Table: Typical Analysis of Coke Oven Gas

Methane	26.0%
Hydrogen	56.5%
Hydrocarbons	2.3%
Carbon monoxide	8.5%
Carbon dioxide	3.0%
Oxygen	0.4%
Nitrogen	3.3%
Density	0.4848 kg/m^3
Calorific value	4300 kcal/m^3

Ammonium Sulphate Plant

The gases from exhaust goes to ESP where tar is separated and the tar free gases goes bubbled through dil. solution of sulphuric acid in saturators. Ammonia is absorbed by sulphuric acid and ammonium sulphate is formed. One tonne of coal yields about 0.3 tonne tar and 5-8 gm ammonia per m^3 of gas.

Benzoyl Recovery Section

The gases from saturator goes to series of coolers and then to benzoyl scrubbers where benzoyl is scrubbed with wash oil. Benzoyl crude oil goes to benzoyl recovery section where benzoyl is removed and the wash oil after treatment is sent to the scrubbers. Crude Benzoyl thus recovered goes to benzoyl rectification plant. Light crude benzoyl contains low boiling sulphur compound, BTX, solvents, still bottom residue. Benzoyl after washing and neutralization with caustic soda is send to benzoyl column for fractionating into different fraction. Various products of coke oven and after distillation of benzoyl are given in Table.

Table : Various products of Coke Oven and Distillation of Benzoyl

Coke (76%)	76%
Tar Productions (3.3%)	(pitch, PCM, anthracene oil, naphthalene, Road tar, cresolate, sodium phenolate, dephenolised oil, other oil)
Ammonia (0.28%)	Used for production of ammonium sulphate
Crude benzoyl (0.85%)	Benzene Toluene, Xylene, Still bottom, Solvent oils
Coke oven gas	For industrial use as fuel
Moisture and other losses	5.04%

Coal Tar Distillation

Coal tar is produced as result of high temperature carbonization and is a viscous dark brown product with characteristic odour and consists of about 300 different products. some of the major constituents are the aromatics and heterocyclic compounds; benzene, toluene, xylene, phenol cresol, naphthalene, anthracene, phenanthrene, pyridine, carbazole, coumarone etc.. Typical composition of coal tar is given in Table.

The tar distillation unit consists of:

- Distillation section
- Fractional crystallization and washing section
- Combustible mixture preparation section
- Phenol rectification section
- dolomite tar preparation unit
- Extra hard pitch preparation

Tar containing around 5% moisture is first dehydrated before distillation. The dehydrated tar is heated to 375-400oC using superheated steam to drive out the flashed vapour and the residue is taken as pitch. The oil vapour is sent to anthracite column for anthracite recovery while the vapour is sent to other column for recovery of various fraction light oil, phenol, naphthalene and heavy oil fraction. Naphthalene fraction is sent to crystalliser to separate naphthalene. Phenol is recovered from various fractions by treating with a sodium hydroxide to form sodium phenolate which is reacted with CO_2 to release phenol. Pyridine is recovered by washing different fraction with sulphuric acid.

Table: Typical Component of Coal Tar

Constituents	Content,%
Naphthalene	5-10
Phenanthrene	4-6
Carbazole	1-2
Anthracene	0.5-1.5
Phenol	0.2-0.5
Crezol	0.6-1.2
Pyridine Compounds	0.5-1.5

Cleaner Technologies in Coke Oven Plant: Coke oven plants are one of the highly polluting industries. Continuous development has been there to reduce the pollution load and energy consumption. Some of the cleaner technology are modified wet quenching, coke dry quenching, coal moisture control,, high pressure ammonia aspiration system, modern leak proof doors, advance technologies for desulphurization of coke oven plant.

Coal or hydrogenation of coal involves raising the atomic hydrogen to carbon ratio. Coal can be converted to liquid and gaseous fuels by direct and indirect processing. Hydrogenation of coal is also called liquefaction of coal. The source of coal is from various coal mines.

Hydrogenation Reaction

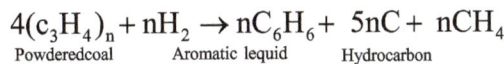

$$4(c_3H_4)_n + nH_2 \rightarrow nC_6H_6 + 5nC + nCH_4$$
$$\text{Powderedcoal} \qquad \text{Aromatic lequid} \qquad \text{Hydrocarbon}$$

This is highly exothermic reaction.

Process flow sheet: Illustrated in Figure.

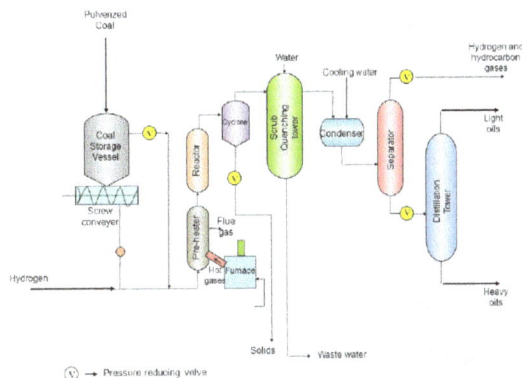

Flow sheet of hydrogenation of coal

Functional Role of each Unit

(a) Coal storage vessel:

- Here pulverized coal is stored.

- Produced coal is fed to the preheater by a screw conveyor.

- Hydrogen gas stream is added to the fed.

- Hydrogen is obtained from any dehydrogenation process or any other hydrogen synthesis ...process.

- Pressure equalizing line is used to maintain pressure of reaction mixture stream with the help ...of valve.

(b) Preheater:

- Powdered coal and hydrogen are the raw material fed to the preheater.

- Optimum temperature is required for reaction so reaction mixture is preheated instead of ....feeding it in reactor directly.

- Heat required for preheating is supplied by furnace.

- In furnace, combustion of fuel gases is occurring and all hot gases from furnace are passed ....from preheater.

(c) Reactor:

- Temperature is maintained at 400-1000°C and pressure of about 500 to 3000psi.

- The residence time is about 1 to 10 minutes.

- ince reaction is highly exothermic, so cooling jackets are used to control temperature of ...reactor.

- Catalyst used is 1% tungsten or molybdenum oxide.

- Catalyst solution is made for reaction.

- The products from the reactor then pass to cyclone.

- There will be liquid aromatics along with unreacted coal and hydrogen in stream leaving the ...reactor.

(d) Cyclone:

- Solids as bottom product are removed.

- Hot reaction mixture from the reactor is fed to the cyclone for separation of solids and gases.

- Gaseous mixture leaving cyclone contain aromatics, unused hydrogen and small amounts of ..light hydrocarbon and carbon dust.

Temperature of the leaving stream is about 700°C.

(e) Scrub quenching tower:

- The gases from cyclone are at very high temperature.
- Sudden cooling of gases or quenching of gases is done in this tower.
- Quenching and absorption both are done simultaneously.
- Water scrub carbon dust from gaseous mixture.
- Stream leaving mainly contain aromatics, hydrogen and other hydrocarbons.

(f) Condenser:

- All the gaseous mixture from the scrubber is fed to the condenser to condense aromatic.liquids.
- The stream leaving the condenser, the aromatics are in liquid form, whereas, hydrogen and other hydrocarbons are in gaseous form.

(g) Separator:

- This unit separate aromatic liquids from other gaseous impurities.
- Hydrogen and other lighter hydrocarbons (gaseous form) are obtained as top product and aromatic.liquids are obtained as bottom product.

(h) Distillation tower:

- Light oil and heavy oils are separated by distillation tower as top and bottom product ....respectively.
- The process occurs at high pressure. So at first, pressure is reduced before distillation by ....pressure reducing valve.
- Reduction in pressure increases relative volatility of heavy oils and light oils.
- Lower temperature and less residence time reduces thermal degradation of aromatic oils.

Coal Production and Consumption

The global coal production in 2011 was 7 billion tones of which China accounted for approximately half of the production and consumption. Total coal production in India during 2009-10 was 532.29 million tones. Lignite production in 2009-10 was 23.95 million tones

Types Of Coal

Coal are classified into various grades based on the composition and calorific value and degree of coalification that has occurred during its formation. Coal may be also classified as hard or soft coal, low sulphur or high sulphur coal. Coal may be also classified in rock types based on petro logical components known as maceral. Based on maceral content coal may be classified as clarain, durain, fusain and vitrain. Classification of different type of coal is mention in Table.

Ultimate analysis of non-coking (Thermal) coal from three power stations (Kahalgaon, Simhadri, and Sipat) is shown in Table along with analysis of Ohio coal in the United States and Long Kou coal from China.

Table : Classification of Different Type of Coal

Types of coal	Description
Peat	Peat is the precursor of coal formed
Lignite	With further increase in temperature during coal formation peat is converted to lignite. Lignite is considered as immature coal. Lignite are brown coloured, soft, low calorific value coal. It is compact in texture.
Sub-bituminous	Sub-bituminous coals are black coloured and are more homogeneous in appearance and their properties range from lignite to that bituminous coal.
Bituminous coal	Bituminous coal is usually black, with higher carbon content and calorific value
Anthracite coal	Anthracite is highest rank coal is a harder, glassy black coal with highest content of carbon and calorific value. Anthracite coal is best suited for making metallurgical coke, for gasification to produce synthesis gas and for combustion as fuel for power generation. The ash content is low.
Graphite	Graphite is the highest rank and is difficult to ignite

Table : Typical Coal Characteristics in selected Indian Power Plants, Compared to selected Chinese and U.S. Coals

Details, %	Kahalgaon	Simhadri	Sipat	US (Ohio)	China (Long Kou)
Carbon	25.07	29.00	30.72	64.2	62.8
Hydrogen	2.95	1.88	2.30	5.0	5.6
Nitrogen	0.50	0.52	0.60	1.3	1.4
Oxygen	6.71	6.96	5.35	11.8	21.7
Moisture	18.5	15.0	15.0	2.8	11.0
Sulphur	0.17	0.25	0.40	1.8	0.9
Ash	46.0	46.0	45.0	16.0	7.7
Calorific Value, kcal/kg	2450	2800	3000	6378	6087

Selection of coal for various applications depends on its composition and carbon content, calorific value, moisture content, ash content, composition of ash, fusion temperature of ash, coking quality, sulphur content.

Assesment of Coal Quality

Coal quality plays an important role in its efficient utilization as fuel and for gasification. It should have high calorific value, high carbon content with low ash content, low sulphur, low moisture, low cost. The quality of coal depends upon it rank. The coal rank is arranged in ascending order

Lignite Sub--bituminous coal--bituminous coal--anthracite

Coal quality can be assessed by proximate, ultimate analysis and calorific value of the coal. Proximate analysis involves determination of moisture, volatile matter, ash and fixed carbon. Ultimate analysis involves determination of carbon and hydrogen,, nitrogen, sulphur, oxygen Calorific value is represented as higher calorific value (HCV) or Gross calorific value (GCV) and Lower calorific value (LCV) or Net calorific value (NCV). Another term used to express energy content is Useful heating value (UHV). UHV is defined as UHV kcal/kg=[8900-138x(percentage of ash content+ percentage of moisture content)]

Coal as Fuel

Coal accounts for 53 percent of the commercial energy sources in India which is high compared to the world average of 30 percent. The 11th plan projected India's coal demand to grow at 975 per annum against 5.7 percent during 10th plan almost two-fold increase. The commercial coal consumed by India 72 percent for power, 14 percent for steel, 9 percent for cement and 9 percent for others. Allocation of coal blocks to private companies are given in Table.

Table : Allocation of Coal Blocks to Private Companies

Sector/End use	Blocks	Geological Reserves(MT)
Power	20	2702
Iron and Steel	47	6703
Small and Isolated	2	9
Cement	3	232
Ultra Mega Power Project	7	2607
Total	**79**	**12254**

Coal as Chemical Feed Stock

Coal originally was utilized as fuel. Many of the petrochemicals now derived from petroleum and natural gas was referred as coal chemicals. With starting of coke oven plants it became source of organic and some inorganic chemicals. Coal tar from coke-oven plants continues to be a source of aromatics, naphthalene and other valuable aromatics like pyridine, picoline, quinolene. Before the coming of petrochemical production a large number of organic chemicals was produced from acetylene produced from calcium carbide route in which coal was a important feed. Various coal chemicals derived from coal is given in Figure.

China has come in a big way for production of chemicals from coal because of the huge coal reserves. With the rising cost of crude oil and dwindling crude oil reserves, coal has again received attention all over the world to utilize coal as an alternative source of chemical feedstock. Various routes for production of organic and inorganic chemicals from coal are

- Coal carbonization and coal tar distillation
- Coal gasification and use of synthesis gas as feed stock for ammonia production
- Coal liquefaction by hydrogenation
- Coal to methanol technology
- Coal to olefin technology
- Coal to plastic technology
- Acetylene from calcium carbide made from lime and coal

Coal Carbonization and Coal Tar Distillation

Coal carbonization and coal tar distillation is integral part of coke oven nits in steel plant for production of coke where large number of chemicals like ammonia, naphtha, aromatics etc are produced benzene, toluene, xylenes were earlier produced from coal tar distillation obtained from coke oven plant.

Coal Based Power Generation

Coal Gasification and Use of Synthesis Gas as Feed Stock for Ammonia Production: Partial oxidation of coal is used for production of synthesis and ammonia. CO from partial oxidation is converted to CO_2 which is used for Urea manufacture. Synthesis gas $CO+H_2$ is used as chemical feedstock for production of large number chemicals.

Coal Liquefaction by Hydrogenation

Coal can be converted to naphtha by direct or indirect liquefaction. Coal can be also converted into naphtha via FT process.

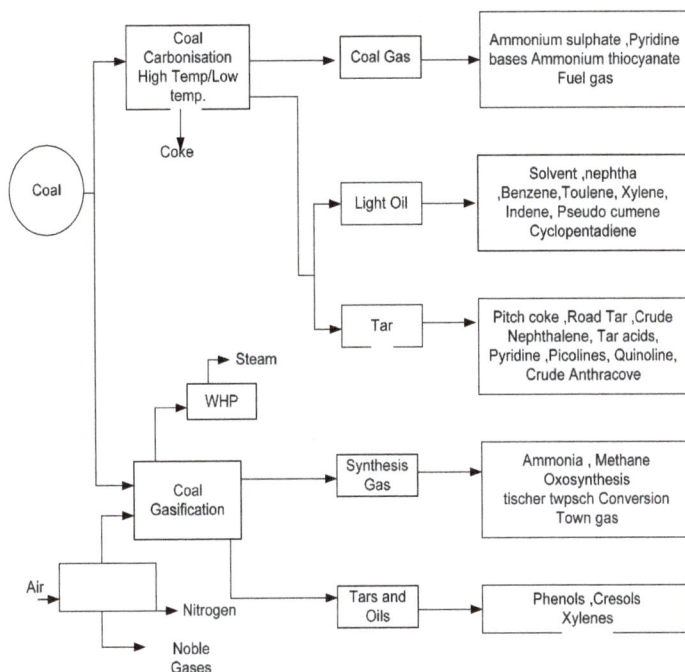

Coal as Chemical Feed Stock

Coal to Methanol Technology

The process involves the production of synthesis gas from coal via partial oxidation followed by conversion of synthesis gas to methanol

Coal to Olefin Technology Methanol can be converted to dimethyl ether, which can be used as a gasoline blend or converted to olefin through olefin synthesis reactions.

Coal to Plastic Technology

Olefin produced from the coal route can be use for the manufacture of plastic. Coal gasification and production of fuel from FT process getting significant attention. Coal gasification and production of synthesis gas is now considered to be a major potential on purpose source of commodity petrochemicals. The cost of coal and higher cost of chemicals derived from coal has been major constraint in its utilization as a substitute for petroleum and natural gas.

Coal Gasification

Coal gasification is the process of producing syngas—a mixture consisting primarily of carbon monoxide (CO), hydrogen (H_2), carbon dioxide (CO_2), methane (CH_4), and water vapor (H_2O)—from coal and water, air and/or oxygen.

Historically, coal was gasified using early technology to produce coal gas (also known as "town gas"), which is a combustible gas traditionally used for municipal lighting and heating before the advent of industrial-scale production of natural gas.

In current practice, large-scale instances of coal gasification are primarily for electricity generation, such as in integrated gasification combined cycle power plants, for production of chemical feedstocks, or for production of synthetic natural gas. The hydrogen obtained from coal gasification can be used for various purposes such as making ammonia, powering a hydrogen economy, or upgrading fossil fuels.

Alternatively, coal-derived syngas can be converted into transportation fuels such as gasoline and diesel through additional treatment via the Fischer-Tropsch process or into methanol which itself can be used as transportation fuel or fuel additive, or which can be converted into gasoline by the methanol to gasoline process. Methane from coal gasification can be converted into LNG for use as a fuel in the transport sector.

History

In the past, coal was converted to make coal gas, which was piped to customers to burn for illumination, heating, and cooking. High prices of oil and natural gas are leading to increased interest in "BTU Conversion" technologies such as gasification, methanation and liquefaction. The Synthetic Fuels Corporation was a U.S. government-funded corporation established in 1980 to create a market for alternatives to imported fossil fuels (such as coal gasification). The corporation was discontinued in 1985.

Early History of Coal gas Production by Carbonization

The Flemish scientist Jan Baptista van Helmont used the name "gas" in his *Origins of Medicine* (c. 1609) to describe his discovery of a "wild spirit" which escaped from heated wood and coal, and which "differed little from the chaos of the ancients". Similar experiments were carried out in 1681 by Johann Becker of Munich and in 1684 by John Clayton of Wigan, England. The latter called it "Spirit of the Coal". William Murdoch (later known as Murdock) discovered new ways of making, purifying and storing gas. Among others, he illuminated his house at Redruth and his cottage at Soho, Birmingham in 1792, the entrance to the Manchester Police Commissioners premises in 1797, the exterior of the factory of Boulton and Watt in Birmingham, and a large cotton mill in Salford, Lancashire in 1805.

Professor Jan Pieter Minckeleers lit his section room at the University of Louvain in 1783 and Lord Dundonald lit his house at Culross, Scotland, in 1787, the gas being carried in sealed vessels from the local tar works. In France, Philippe le Bon patented a gas fire in 1799 and demonstrated street lighting in 1801. Other demonstrations followed in France and in the United States, but, it is generally recognized that the first commercial gas works was built by the London and Westminster Gas Light and Coke Company in

Great Peter Street in 1812 laying wooden pipes to illuminate Westminster Bridge with gas lights on New Year's Eve in 1813. In 1816, Rembrandt Peale and four others established the Gas Light Company of Baltimore, the first manufactured gas company in America. In 1821, natural gas was being used commercially in Fredonia, New York. The first German gas works was built in Hannover in 1825 and by 1870 there were 340 gas works in Germany making town gas from coal, wood, peat and other materials.

Working conditions in the Gas Light and Coke Company's Horseferry Road Works, London, in the 1830s were described by a French visitor, Flora Tristan, in her *Promenades Dans Londres*:

Two rows of furnaces on each side were fired up; the effect was not unlike the description of Vulcan's forge, except that the Cyclops were animated with a divine spark, whereas the dusky servants of the English furnaces were joyless, silent and benumbed. The foreman told me that stokers were selected from among the strongest, but that nevertheless they all became consumptive after seven or eight years of toil and died of pulmonary consumption. That explained the sadness and apathy in the faces and every movement of the hapless men.

The first public piped gas supply was to 13 gas lamps, each with three glass globes along the length of Pall Mall, London in 1807. The credit for this goes to the inventor and entrepreneur Fredrick Winsor and the plumber Thomas Sugg, who made and laid the pipes. Digging up streets to lay pipes required legislation and this delayed the development of street lighting and gas for domestic use. Meanwhile, William Murdoch and his pupil Samuel Clegg were installing gas lighting in factories and work places, encountering no such impediments.

Early History of Coal Gas Production by Gasification

In the 1850s every small to medium-sized town and city had a gas plant to provide for street lighting. Subscribing customers could also have piped lines to their houses. By this era, gas lighting became accepted. Gaslight trickled down to the middle class and later came gas cookers and stoves.

The 1860s were the golden age of coal gas development. Scientists like Kekulé and Perkin cracked the secrets of organic chemistry to reveal how gas is made and its composition. From this came better gas plants and Perkin's purple dyes, such as Mauveine. In the 1850s, processes for making Producer gas and Water gas from coke were developed. Unenriched water gas may be described as Blue water gas (BWG).

Mond gas, developed in the 1850s by Ludwig Mond, was producer gas made from coal instead of coke. It contained ammonia and coal tar and was processed to recover these valuable compounds.

Blue water gas (BWG) burns with a non-luminous flame which makes it unsuitable for lighting purposes. Carburetted Water Gas (CWG), developed in the 1860s, is BWG enriched with gases obtained by spraying oil into a hot retort. It has a higher calorific value and burns with a luminous flame.

The carburetted water gas process was improved by Thaddeus S. C. Lowe in 1875. The gas oil was fixed into the BWG via thermocracking in the carburettor and superheater of the CWG generating set. CWG was the dominant technology in the USA from the 1880s until the 1950s, replacing coal gasification. CWG has a CV of 20 MJ/m^3 i.e. slightly more than half that of natural gas.

Development of the Coal Gas Industry in the UK

The advent of incandescent gas lighting in factories, homes and in the streets, replacing oil lamps and candles with steady clear light, almost matching daylight in its colour, turned night into day for many—making night shift work possible in industries where light was all important—in spinning, weaving and making up garments etc. The social significance of this change is difficult for generations brought up with lighting after dark available at the touch of a switch to appreciate. Not only was industrial production accelerated, but streets were made safe, social intercourse facilitated and reading and writing made more widespread. Gas works were built in almost every town, main streets were brightly illuminated and gas was piped in the streets to the majority of urban households. The invention of the gas meter and the pre-payment meter in the late 1880s played an important role in selling town gas to domestic and commercial customers.

The education and training of the large workforce, the attempts to standardise manufacturing and commercial practices and the moderating of commercial rivalry between supply companies prompted the founding of associations of gas managers, first in Scotland in 1861. A British Association of Gas Managers was formed in 1863 in Manchester and this, after a turbulent history, became the foundation of the Institute of Gas Engineers (IGE). In 1903, the reconstructed Institution of Civil Engineers (ICE) initiated courses for students of gas manufacture in the City and Guilds of London Institute. The IGE was granted the Royal Charter in 1929. Universities were slow to respond to the needs of the industry and it was not until 1908 that the first Professorship of Coal Gas and Fuel Industries was founded at the University of Leeds. In 1926, the Gas Light and Coke Company opened *Watson House* adjacent to Nine Elms Gas Works. At first, this was a scientific laboratory. Later it included a centre for training apprentices but its major contribution to the industry was its gas appliance testing facilities, which were made available to the whole industry, including gas appliance manufacturers. Using this facility, the industry established not only safety but also performance standards for both the manufacture of gas appliances and their servicing in customers' homes and commercial premises.

During World War I, the gas industry's by-products, phenol, toluene and ammonia and sulphurous compounds were valuable ingredients for explosives. Much coal for the gas works was shipped by sea and was vulnerable to enemy attack. The gas industry was a large employer of clerks, mainly male before the war. But the advent of the typewriter and the female typist made another important social change that was, unlike the employment of women in war-time industry, to have long-lasting effects.

The inter-war years were marked by the development of the continuous vertical retort which replaced many of the batch fed horizontal retorts. There were improvements in storage, especially the waterless gas holder, and distribution with the advent of 2–4 inch steel pipes to convey gas at up to 50 psi (340 kPa) as feeder mains compared to the traditional cast iron pipes working at an average of 2–3 inches water gauge (500–750 Pa). Benzole as a vehicle fuel and coal tar as the main feedstock for the emerging organic chemical industry provided the gas industry with substantial revenues. Petroleum supplanted coal tar as the primary feedstock of the organic chemical industry after World War II and the loss of this market contributed to the economic problems of the gas industry after the war.

A wide variety of appliances and uses for gas developed over the years. Gas fires, gas cookers, refrigerators, washing machines, hand irons, pokers for lighting coal fires, gas-heated baths, remotely controlled clusters of gas lights, gas engines of various types and, in later years, gas warm air and hot water central heating and air conditioning, all of which made immense contributions to the improvement of the quality of life in cities and towns worldwide. The evolution of electric lighting made available from public supply extinguished the gas light, except where colour matching was practised as in haberdashery shops.

Process

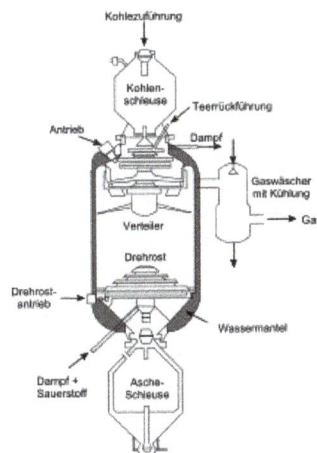

Scheme of a Lurgi gasifier

During gasification, the coal is blown through with oxygen and steam (water vapor) while also being heated (and in some cases pressurized). If the coal is heated by external heat sources the process is called "allothermal", while "autothermal" process assumes heating of the coal via exothermal chemical reactions occurring inside the gasifier itself. It is essential that the oxidizer supplied is insufficient for complete oxidizing (combustion) of the fuel. During the reactions mentioned, oxygen and water molecules oxidize the coal and produce a gaseous mixture of carbon dioxide (CO_2), carbon monoxide (CO), water vapour (H_2O), and molecular hydrogen (H_2). (Some by-products like tar, phenols, etc. are also possible end products, depending on the specific gasification technology utilized.) This process has been conducted in-situ within natural coal seams (referred to as underground coal gasification) and in coal refineries. The desired end product is usually syngas (i.e., a combination of H_2 + CO), but the produced coal gas may also be further refined to produce additional quantities of H_2:

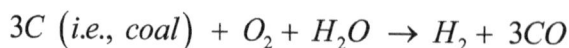

$$3C \left(i.e.,\ coal\right) + O_2 + H_2O \rightarrow H_2 + 3CO$$

If the refiner wants to produce alkanes (i.e., hydrocarbons present in natural gas, gasoline, and diesel fuel), the coal gas is collected at this state and routed to a Fischer-Tropsch reactor. If, however, hydrogen is the desired end-product, the coal gas (primarily the CO product) undergoes the water gas shift reaction where more hydrogen is produced by additional reaction with water vapor:

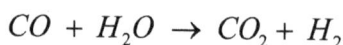

$$CO + H_2O \rightarrow CO_2 + H_2$$

Although other technologies for coal gasification currently exist, all employ, in general, the same chemical processes. For low-grade coals (i.e., "brown coals") which contain significant amounts of water, there are technologies in which no steam is required during the reaction, with coal (carbon) and oxygen being the only reactants. As well, some coal gasification technologies do not require high pressures. Some utilize pulverized coal as fuel while others work with relatively large fractions of coal. Gasification technologies also vary in the way the blowing is supplied.

"Direct blowing" assumes the coal and the oxidizer being supplied towards each other from the opposite sides of the reactor channel. In this case the oxidizer passes through coke and (more likely) ashes to the reaction zone where it interacts with coal. The hot gas produced then passes fresh fuel and heats it while absorbing some products of thermal destruction of the fuel, such as tars and phenols. Thus, the gas requires significant refining before being used in the Fischer-Tropsch reaction. Products of the refinement are highly toxic and require special facilities for their utilization. As a result, the plant utilizing the described technologies has to be very large to be economically efficient. One of such plants called SASOL is situated in the Republic of South Africa (RSA). It was built due to embargo applied to the country preventing it from importing oil and natural gas. RSA is rich in Bituminous coal and Anthracite and was able to arrange the use of the well known high pressure "Lurgi" gasification process developed in Germany in the first half of 20-th century.

"Reversed blowing" (as compared to the previous type described which was invented first) assumes the coal and the oxidizer being supplied from the same side of the reactor. In this case there is no chemical interaction between coal and oxidizer before the reaction zone. The gas produced in the reaction zone passes solid products of gasification (coke and ashes), and CO_2 and H_2O contained in the gas are additionally chemically restored to CO and H_2. As compared to the "direct blowing" technology, no toxic by-products are present in the gas: those are disabled in the reaction zone. This type of gasification has been developed in the first half of 20-th century, along with the "direct blowing", but the rate of gas production in it is significantly lower than that in "direct blowing" and there were no further efforts of developing the "reversed blowing" processes until 1980-s when a Soviet research facility KATEKNIIUgol' (R&D Institute for developing Kansk-Achinsk coal field) began R&D activities to produce the technology now known as "TERMOKOKS-S" process. The reason for reviving the interest to this type of gasification process is that it is ecologically clean and able to produce two types of useful products (simultaneously or separately): gas (either combustible or syngas) and middle-temperature coke. The former may be used as a fuel for gas boilers and diesel-generators or as syngas for producing gasoline, etc., the latter - as a technological fuel in metallurgy, as a chemical absorbent or as raw material for household fuel briquettes. Combustion of the product gas in gas boilers is ecologically cleaner than combustion of initial coal. Thus, a plant utilizing gasification technology with the "reversed blowing" is able to produce two valuable products of which one has relatively zero production cost since the latter is covered by competitive market price of the other. As the Soviet Union and its KATEKNIIUgol' ceased to exist, the technology was adopted by the individual scientists who originally developed it and is now being further researched in Russia and commercially distributed worldwide. Industrial plants utilizing it are now known to function in Ulaan-Baatar (Mongolia) and Krasnoyarsk (Russia).

Pressurized airflow bed gasification technology created through the joint development between Wison Group and Shell (Hybrid). For example: Hybrid is an advanced pulverized coal gasification technology, this technology combined with the existing advantages of Shell SCGP waste heat boiler, includes more than just a conveying system, pulverized coal pressurized gasification burner arrangement, lateral jet burner membrane type water wall, and the intermittent discharge has been fully validated in the existing SCGP plant such as mature and reliable technology, at the same time, it removed the existing process complications and in the syngas cooler (waste pan) and [fly ash] filters which easily failed, and combined the current existing gasification technology that is widely used in synthetic gas quench process. It not only retains the original Shell SCGP waste heat boiler of coal characteristics of strong adaptability, and ability to scale up easily, but also absorb the advantages of the existing quench technology.

Underground Coal Gasification

Underground coal gasification is an industrial gasification process, which is carried out in non-mined coal seams using injection of a gaseous oxidizing agent, usually oxygen or air, and bringing the resulting product gas to surface through production wells drilled from the surface. The product gas could to be used as a chemical feedstock or as fuel for power generation. The technique can be applied to resources that are otherwise not economical to extract and also offers an alternative to conventional coal mining methods for some resources. Compared to traditional coal mining and gasification, UCG has less environmental and social impact, though some concerns including potential for aquifer contamination are known.

Carbon Capture Technology

Carbon capture, utilization, and sequestration (or storage) is increasingly being utilized in modern coal gasification projects to address the greenhouse gas emissions concern associated with the use of coal and carbonaceous fuels. In this respect, gasification has a significant advantage over conventional coal combustion, in which CO_2 resulting from combustion is considerably diluted by nitrogen and residual oxygen in the near-ambient pressure combustion exhaust, making it relatively difficult, energy-intensive, and expensive to capture the CO_2 (this is known as "post-combustion" CO_2 capture).

In gasification, on the other hand, oxygen is normally supplied to the gasifiers and just enough fuel is combusted to provide the heat to gasify the rest; moreover, gasification is often performed at elevated pressure. The resulting syngas is typically at higher pressure and not diluted by nitrogen, allowing for much easier, efficient, and less costly removal of CO_2. Gasification and integrated gasification combined cycle's unique ability to easily remove CO_2 from the syngas prior to its combustion in a gas turbine (called "pre-combustion" CO_2 capture) or its use in fuels or chemicals synthesis is one of its significant advantages over conventional coal utilization systems.

CO_2 Capture Technology Options

All coal gasification-based conversion processes require removal of hydrogen sulfide (H_2S; an acid gas) from the syngas as part of the overall plant configuration. Typical acid gas removal (AGR) processes employed for gasification design are either a chemical solvent system (e.g., amine gas treating systems based on MDEA, for example) or a physical solvent system (e.g., Rectisol or Selexol). Process selection is mostly dependent on the syngas cleanup requirement and costs. Conventional chemical/physical AGR processes using MDEA, Rectisol or Selexol are commercially proven technologies and can be designed for selective removal of CO_2 in addition to H_2S from a syngas stream. For significant capture of CO_2 from a gasification plant (e.g., > 80%) the CO in the syngas must first be converted to CO_2 and hydrogen (H_2) via a water-gas-shift (WGS) step upstream of the AGR plant.

For gasification applications, or IGCC, the plant modifications required to add the ability to capture CO_2 are minimal. The syngas produced by the gasifiers needs to be treated through various processes for the removal of impurities already in the gas stream, so all that is required to remove CO_2 is to add the necessary equipment, an absorber and regenerator, to this process train. In combustion applications, modifications must be done to the exhaust stack and because of the lower concentrations of CO_2 present in the exhaust, much larger volumes of total gas require processing, necessitating larger and more expensive equipment.

IGCC-based Projects in the United States with CO_2 Capture and use/Storage

Mississippi Power's Kemper Project is in late stages of construction. It will be a lignite-fuel IGCC plant, generating a net 524 MW of power from syngas, while capturing over 65% of CO_2 generated using the Selexol process. The technology at the Kemper facility, Transport-Integrated Gasification (TRIG), was developed and is licensed by KBR. The CO_2 will be sent by pipeline to depleted oil fields in Mississippi for enhanced oil recovery operations.

Hydrogen Energy California (HECA) will be a 300MW net, coal and petroleum coke-fueled IGCC polygeneration plant (producing hydrogen for both power generation and fertilizer manufacture). Ninety percent of the CO_2 produced will be captured (using Rectisol) and transported to Elk Hills Oil Field for EOR, enabling recovery of 5 million additional barrels of domestic oil per year.

Summit's Texas Clean Energy Project (TCEP) will be a coal-fueled, IGCC-based 400MW power/polygeneration project (also producing urea fertilizer), which will capture 90% of its CO_2 in pre-combustion capture using the Rectisol process. The CO_2 not used in fertilizer manufacture will be used for enhanced oil recovery in the West Texas Permian Basin.

Plants such as the Texas Clean Energy Project which employ carbon capture and storage have been touted as a partial, or interim, solution to climate change issues if they can be made economically viable by improved design and mass production. There was opposition by utility regulators and ratepayers due to increased cost and by some environmentalists such as Bill McKibben who view any continued use of fossil fuels as counterproductive.

By-products

The by-products of coal gas manufacture included coke, coal tar, sulfur and ammonia; all useful products. Dyes, medicines, including sulfa drugs, saccharin and many organic compounds are therefore derived from coal gas.

Coke is used as a smokeless fuel and for the manufacture of water gas and producer gas. Coal tar is subjected to fractional distillation to recover various products, including

- tar, for roads

- benzole, a motor fuel

- creosote, a wood preservative

- phenol, used in the manufacture of plastics

- cresols, disinfectants

- Sulfur is used in the manufacture of sulfuric acid and ammonia is used in the manufacture of fertilisers.

Commercialization

According to the Gasification and Syngas Technologies Council, a trade association, there are globally 272 operating gasification plants with 686 gasifiers and 74 plants with 238 gasifiers under construction. Most of them use coal as feedstock.

As of 2017 large scale expansion of the coal gasification industry was occurring only in China where local governments and energy companies promote the industry for the sake of jobs and a market for coal. The central government is aware of the conflict with environmental goals. For the most part the plants are located in remote coal rich areas. In addition to producing a great deal of carbon dioxide the plants use a great deal of water in areas where water is scarce.

Environmental Impact

Environmental Impact of Manufactured Coal Gas Industry

From its original development until the wide-scale adoption of natural gas, more than 50,000 manufactured gas plants were in existence in the United States alone. The process of manufacturing gas usually produced a number of by-products that contaminated the soil and groundwater in and around the manufacturing plant, so many former town gas plants are a serious environmental concern, and cleanup and remediation costs are often high. Manufactured gas plants (MGPs) were typically sited near or adjacent to waterways that were used to transport in coal and for the discharge of wastewater contaminated with tar, ammonia and/or drip oils, as well as outright waste tars and tar-water emulsions.

In the earliest days of MGP operations, coal tar was considered a waste and often disposed into the environment in and around the plant locations. While uses for coal tar developed by the late-19th century, the market for tar varied and plants that could not sell tar at a given time could store tar for future use, attempt to burn it as fuel for the boilers, or dump the tar as waste. Commonly, waste tars were disposed of in old gas holders, adits or even mine shafts (if present). Over time, the waste tars degrade with phenols, benzene (and other mono-aromatics – BTEX) and polycyclic aromatic hydrocarbons released as pollutant plumes that can escape into the surrounding environment. Other wastes included "blue billy", which is a ferroferricyanide compound—the blue colour is from Prussian blue, which was commercially used as a dye. Blue billy is typically a granular material and was sometimes sold locally with the strap line "guaranteed weed free drives". The presence of blue billy can give gas works waste a characteristic musty/bitter almonds or marzipan smell which is associated with cyanide gas.

The shift to the CWG process initially resulted in a reduced output of water gas tar as compared to the volume of coal tars. The advent of automobiles reduced the availability of naphtha for carburetion oil, as that fraction was desirable as motor fuel. MGPs that shifted to heavier grades of oil often experienced problems with the production of tar-water emulsions, which were difficult, time consuming, and costly to break. (The cause of tar change water emulsions is complex and was related to several factors, including free carbon in the carburetion oil and the substitution of bituminous coal as a feedstock instead of coke.) The production of large volumes of tar-water emulsions quickly filled up available storage capacity at MGPs and plant management often dumped the emulsions in pits, from which they may or may not have been later reclaimed. Even if the emulsions were reclaimed, the environmental damage from placing tars in unlined pits remained. The dumping of emulsions (and other tarry residues such as tar sludges, tank bottoms, and off-spec tars) into the soil and waters around MGPs is a significant factor in the pollution found at FMGPs today.

Commonly associated with former manufactured gas plants (known as "FMGPs" in environmental remediation) are contaminants including:

- BTEX
  - Diffused out from deposits of coal/gas tars
  - Leaks of carburetting oil/light oil
  - Leaks from drip pots, that collected condensible hydrocarbons from the gas
- Coal tar waste/sludge
  - Typically found in sumps of gas holders and decanting ponds.
  - Coal tar sludge has no resale value and so was always dumped.
- Volatile organic compounds
- Polycyclic aromatic hydrocarbons (PAHs)
  - Present in coal tar, gas tar, and pitch at significant concentrations.
- Heavy metals
  - Leaded solder for gas mains, lead piping, coal ashes.
- Cyanide
  - Purifier waste has large amounts of complex ferrocyanides in it.

- Lampblack

  - Only found where crude oil was used as gasification feedstock.

- Tar emulsions

Coal tar and coal tar sludges are frequently denser than water and are present in the environment as a dense non-aqueous phase liquid.

In the UK, former gasworks have commonly been developed over for residential and other uses (including the Millennium Dome), being seen as prime developable land in the confines of city boundaries. Situations such as these are now leading to problems associated with planning and the Contaminated Land Regime and have recently been debated in the House of Commons.

Environmental Impact of Modern Coal Gasification

Coal gasification processes require controls and pollution prevention measures to mitigate pollutant emissions.Pollutants or emissions of concern in the context of coal gasification include primarily:

- Ash & slag

Non-slagging gasifiers produce dry ash similar to that produced by conventional coal combustion, which can be an environmental liability if the ash (typically containing heavy metals) is leachable or caustic, and if the ash must be stored in ash ponds. Slagging gasifiers, which are utilized at many of the major coal gasification applications worldwide, have considerable advantage in that ash components are fused into a glassy slag, capturing trace heavy metals in the non-leachable glassy matrix, rendering the material non-toxic. This non-hazardous slag has multiple beneficial uses such as aggregate in concrete, aggregate in asphalt for road construction, grit in abrasive blasting, roofing granules, etc.

- Carbon dioxide (CO_2)

CO_2 is of paramount importance in global climate change.

- Mercury

- Arsenic

- Particulate matter (PM)

Ash is formed in gasification from inorganic impurities in the coal. Some of these impurities react to form microscopic solids which can be suspended in the syngas produced by gasification.

- Sulfur dioxide (SO_2)

Typically coal contains anywhere from 0.2 to 5 percent sulfur by dry weight, which converts to H_2S and COS in the gasifiers due to the high temperatures and low oxygen levels. These "acid gases" are removed from the syngas produced by the gasifiers by acid gas removal equipment prior to the syngas being burned in the gas turbine to produce electricity, or prior to its use in fuels synthesis.

- Nitrogen oxides (NO_x)

(NO_x) refers to nitric oxide (NO) and nitrogen dioxide (NO_2). Coal usually contains between 0.5 and 3 percent nitrogen on a dry weight basis, most of which converts to harmless nitrogen gas. Small levels of ammonia and hydrogen cyanide are produced, and must be removed during the syngas cooling process. In the case of power generation, NO_x also can be formed downstream by the combustion of syngas in turbines.

Process Steps in Gasification

Various process steps involved in gasification are

Feed Preparation: Wet feed system or dry feed system may be used in gasification. Wet feed system employ grinding and slurring with water and pumping to gasifier while dry system consist of grinding of the feed and employ lock hopers to pressurize the feed

Gasification: Gasification is carried out at temperature between 900-1100°C. carbon react with oxygen and steam to raw synthesis gas(CO and H_2) and some minor by products which is removed to produce clean synthesis gas which can be used as fuel , for generation of steam and electricity or may be used as chemical feed stock.

Syn gas cooling: High temperature syn gas produced from gasification is cooled

Air Separation Unit: The process produces oxygen for gasification. The process is based on cryogenic separation of air. The process involves liquefaction of air and fractionation to get oxygen, nitrogen and other gases.

Acid Gas Removal: This involves removal of impurities like H_2S, COS, NH_3 and HCN by absorption using various commercial solvent like mono-ethanol amine (MEA), methyl diethanol amine (MDEA), methanol rectisolvent and Selexol [Handa and ganesh, 2011]. The solvent is regenerated by stripping the acid gases and recycle.

Slag Handling System: The gasifier may be either slagging or non slagging.

Gasification in Petroleum refinery [Handa and Ganesh,2011]: With increasing use of heavier crude and processing of heavy residue has resulted in increased production of tar, petroccoke and asphalt by various residues upgrading technologies processes like visbreaking, coking and deasphalting. These residues can be gasified for production of hydrogen, syn. gas, electricity ammonia and chemicals. During recent years petrocoke gasification is getting interest by petroleum refiners for production of syn. gas which can be further process to separate hydrogen or can be used for production of useful chemicals.

Gasifier Configurations

Various types of gasifier used are mention in Table and shown in Figure below:

Table : Various Type of Gasifier

Fixed bed	Fixed bed involves an upward flow of reaction gas through a relatively stationary bed of hot coal. The gas velocity is low.
Moving bed gasifier	Moving bed gasifier operates counter currently where the coal inter the gasifier at the top and moves downward and slowly heated with the product gas. Gasification takes place in the gasification zone.
Fluidised bed gasifier e.g., Winkler/ KBR/U- GAS	Fluidised bed operates at higher gas velocities than fixed bed and uses smaller particle. The gasifier operates at atmospheric pressure and moderate and uniform temperature.
Entrained flow gasifier	Entrained bed operates with parallel flows of reaction gas and pulverized coal which minimize the reaction time and maximize throughput of product.. The residence time is few seconds. Ash is removed as molten slag

Gasification Reactions:

$$C_n H_m + n/2O_2 \Rightarrow nCO + m/2H_2 \ (1 < n < 4)$$

Where for gas as pure methane m=4 and n=1 hence m/n=4

For oil m/n≈2 hence m=2, n=1

For coal m/n≈1 hence m=1,n=1

It is exothermic and produces a gas containing mainly CO and H_2. The raw synthesis gas contains small quantities of CO_2, H_2O and H_2S and impurities, such as CH_4, NH_3, COS, HCN, N_2, Argon and ash, the quantities being determined by the composition of the feedstock, the oxidant and actual gasification temperature (1300-1400°C). A small amount of unconverted carbon is also present and ranges from 0.5 to 1.0 by percent wt in liquid feedstock or 50-200 ppm wt in gaseous feedstock. The various reactions can be summarized as follows [Chemical Industry Digest. Annual-January 2010.

Combustion

$$C + 1/2O_2 \Rightarrow CO, \Delta H\ -111MJ/Kmol$$

$$CO + 1/2O_2 \Rightarrow CO_2, \Delta H\ -283MJ/Kmol$$

$$H_2 + 1/2O_2 \Rightarrow H_2O, \Delta H\ -242MJ/Kmol$$

Steam Reforming Reaction

$$CH_4 + H_2O \Rightarrow CO_2 + 3H_2,\ \Delta H\ +206\ MJ\ Kmol$$

Boudourard Reaction

$$C + CO_2 \Leftrightarrow 2CO, \ \Delta H \ +172 \ MJ/Kmol$$

Water Gas Reaction

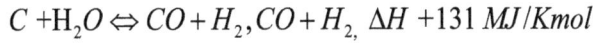

$$C + H_2O \Leftrightarrow CO + H_2, CO + H_2, \ \Delta H \ +131 \ MJ/Kmol$$

Water Shift Reaction

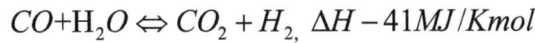

$$CO + H_2O \Leftrightarrow CO_2 + H_2, \ \Delta H \ -41 MJ/Kmol$$

Methanation Reactions

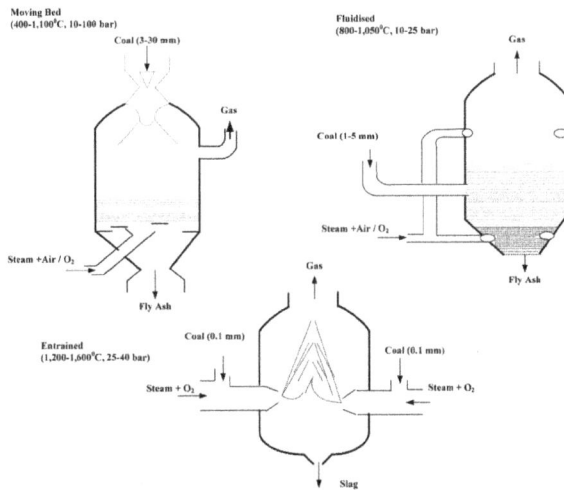

$$C + 2H_2 \Leftrightarrow CH_4, \ \Delta H \ -75 MJ/Kmol$$

Various Type of Gasifier Commonly Used

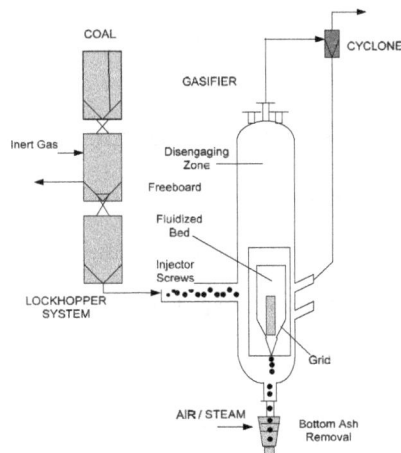

Fluidised bed gasifier

New Gasifier Designs [Parthasarthi,2009]

GE Global/ Unmixed Fuel Processor (UOP):

- Elimination of Air Separation Unit (ASU)

- High Temperature Syngas Clean up

- Higher efficiency

- Lower cost

KBR Transport Gasifier (TRIG™) [Parthasarthi,2009]

- Low rank, high-ash, high-moisture coal compatible

- For power generation, air can be employed as the oxidant

- Lower cost predicted

- Higher availability predicted

- Non-slagging, and refractory issues should therefore be minimal

- Higher predicted efficiency

- Lower emissions (due to higher efficiency)

- Large scale up of the technology still required, by a factor of ~30

- By 2010, this technology will be operating at the scale of E-Gas which has >20 years experience already in 2007.

- Lower temperatures and short gas residence time may lead to some methane formation, which is detrimental in chemical applications

- Ash disposal problem if carbon conversion predictions are not met in commercial apparatus.

Catalytic Coal Gasification – Bluegas™ {Parthasarthi,2009}

- Elimination of oxygen plant

- For SNG objective, little or no catalytic methanation required

- High thermodynamic efficiency potential

- Catalyst cost and recoverability

- Carbon conversion and methane production yields in the gasifier

- Cost of applying the catalyst effectively to the coal

- Inherently must be done in a fluidized bed which have not been scaled up to larger capacities of entrained gasifier (yet)

- Interactions of catalyst with coal ash

- Separation costs of syngas and methane – cryogenic process

- Excess steam requirements

- Unsuitable for chemical synthesis processes due to CH_4 reforming requirement.

Coal Gasification in fertilizer Industry: Coal gasification fro production of synthesis gas and ammonia has being used many part of the world where large quantity of coal is present. However, higher energy consumption in coal based ammonia plant has been major constrain.

Relative energy consumption per tonne of ammonia is given below [Murthy and Patnaik, 2001].

Gas	Naphtha	Fuel oil	Coal
	1.1	1.15	1.45

There has been extensive research for improvement of coal gasification technology to improve the efficiency and reduce the cost.

Rapid Thermal Processing (RTP ™) Process for Conversion of Biomass to Liquid Fuelshg

RTP™ is fully heat integrated technology that yields over 70percent liquid products from typical biomass feed stocks. It is a fast thermal process where biomass is rapidly of heated in the absence of oxygen. The biomass is vaporized and the vapour cooled to generate high yields of pyrolysis oil. Typical yield of various biomass is given in Table. Chemical and bio- chemical conversion of biomass to biofuel is shown in Figures respectively.

Table : Typical Yield of Various Biomass

Biomass Feed stock	Typical Pyrolysis oil yield ,wt% of dry feed stock70-75
Hardwood	70-80
Softwood	70-80
Hardwood bark	60-65
Soft wood bark	55-65
Corn fiber	65-75
Bagasse	70-75
Waste Paper	60-80

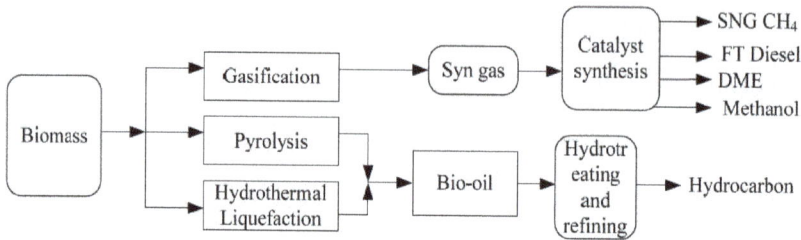

Chemical Conversion Route of Biomass to Biofuels

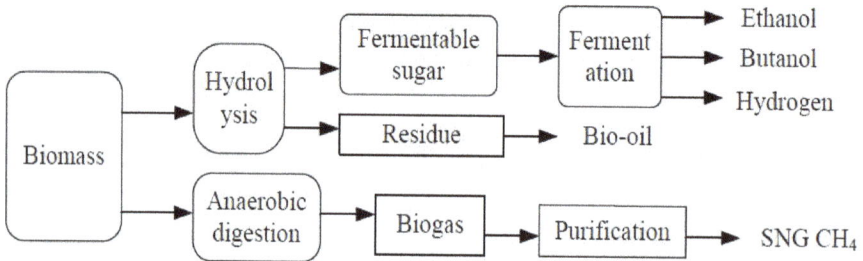

Biochemical conversion route of biomass to biofuels

Co-gasification of Coal and biomass for methanol synthesis: Co- gasification of coal and biomass combined with power and methanol production can be of considered as a potential fuel-base for gasification and further production of synthesis gas and methanol. Some of the advantage of integrated biomass-coal cogasification system is possible continuous operation of the coal gasification reactor, lower costs of electric-energy production [Chemielniak & Sciazko, 2003]

References

- Speight, James G. (2007). Natural Gas: A Basic Handbook. Elsevier. pp. 120–121. ISBN 9780127999845

- Tristan, Flora (1840) Promenades Dans Londres. Trans. Palmer, D, and Pincetl, G. (1980) Flora Tristan's London Journal, A Survey of London Life in the 1830s George Prior, Publishers, London. Extract Worse than the slave trade in Appendix 1, Barty-King, H (1985)

- Everard, Stirling (1949). The History of the Gas Light and Coke Company 1812-1949. London: Ernest Benn Limited. (Reprinted 1992, London: A&C Black (Publishers) Limited for the London Gas Museum. ISBN 0-7136-3664-5) Chapter XX, Sir David Milne-Watson, Bart.: I. Expansion

- Joe Nocera (March 15, 2013). "A Real Carbon Solution" (op-ed based on facts). The New York Times. Retrieved March 16, 2013

- Edward Wong (February 8, 2017). "'Irrational' Coal Plants May Hamper China's Climate Change Efforts". The New York Times. Retrieved February 8, 2017

Petroleum Distillation: A Refining Process

Crude oil is a multicomponent mixture that needs to be refined to obtain valuable products such as diesel oil, liquefied petroleum gas, petrol, etc. The processes used in refining include distillation, alkylation, isomerization, cracking, etc. The following chapter elucidates the varied processes and mechanisms associated with this area of study.

Refining Petroleum

In this section, we present a brief overview of the petroleum refining, a prominent process technology in process engineering.

Crude Oil

Crude oil is a multicomponent mixture consisting of more than 108 compounds. Petroleum refining refers to the separation as well as reactive processes to yield various valuable products. Therefore, a key issue in the petroleum refining is to deal with multicomponent feed streams and multicomponent product streams. Usually, in chemical plants, we encounter streams not possessing more than 10 components, which is not the case in petroleum refining. Therefore, characterization of both crude, intermediate product and final product streams is very important to understand the processing operations effectively.

Overview of Refinery Processes

Primary crude oil cuts in a typical refinery include gases, light/heavy naphtha, kerosene, light gas oil, heavy gas oil and residue. From these intermediate refinery product streams several final product streams such as fuel gas, liquefied petroleum gas (LPG), gasoline, jet fuel, kerosene, auto diesel, lubricants, bunker oil, asphalt and coke are obtained. The entire refinery technology involves careful manipulation of various feed properties using both chemical and physical changes.

Conceptually, a process refinery can be viewed upon as a combination of both physical and chemical processes or unit operations and unit processes respectively. Typically, the dominant physical process in a refinery is the distillation process that enables the removal of lighter components from the heavier components. Other chemical processes such as alkylation and isomerisation are equally important in the refinery engineering as these processes enable the reactive transformation of various functional groups to desired functional groups in the product streams.

Feed and Product Characterization

The characterization of petroleum process streams is approached from both chemistry and physical properties perspective. The chemistry perspective indicates to characterize the crude oil in terms of the functional groups such as olefins, paraffins, naphthenes, aromatics and resins. The dominance of one or more of the functional groups in various petroleum processing streams is indicative of the desired product quality and characterization. For instance, the lighter fractions of the refinery consist of only olefins and paraffins. On the other hand, products such as petrol should have high octane number which is a characteristic feature of olefinic and aromatic functional groups present in the product stream.

The physical characterization of the crude oil in terms of viscosity, density, boiling point curves is equally important. These properties are also indicative of the quality of the product as well as the feed. Therefore, in petroleum processing, obtaining any intermediate or a product stream with a defined characterization of several properties indicates whether it is diesel or petrol or any other product. This is the most important characteristic feature of petroleum processing sector in contrary to the chemical process sector.

The product characterization is illustrated now with an example. Aviation gasoline is characterized using ASTM distillation. The specified temperatures for vol% distilled at 1 atm. Are 158oF maximum for 10 % volume, 221oF maximum for 50 % volume and 275oF maximum for 90% volume. This is indicative of the fact that any product obtained in the refinery process and meets these ASTM distillation characteristics is anticipated to represent Aviation gasoline product. However, other important properties such as viscosity, density, aniline product, sulphur density are as well measured to fit within a specified range and to conclude that the produced stream is indeed aviation gasoline.

Important Characterization Properties

Numerous important feed and product characterization properties in refinery engineering include

API Gravity

API gravity of petroleum fractions is a measure of density of the stream. Usually measured at 60°F, the API gravity is expressed as

$^\circ API = [\ 141.5/\text{specific gravity}\ ] - 131.5$

where specific gravity is measured at 60° F.

According to the above expression, 10° API gravity indicates a specific gravity of 1 (equivalent to water specific gravity). In other words, higher values of API gravity indicate lower specific gravity and therefore lighter crude oils or refinery products and vice-versa. As far as crude oil is concerned, lighter API gravity value is desired as more amount of gas fraction, naphtha and gas oils can be produced from the lighter crude oil than with the heavier crude oil. Therefore, crude oil with high values of API gravity are expensive to procure due to their quality.

Watson Characterization Factor

The Watson characterization factor is usually expressed as

$$k = \left(T_B\right)^{1/3} / _{specific\ gravity}$$

Where T_B is the average boiling point in degrees R taken from five temperatures corresponding to 10, 30, 50, 70 and 90 volume % vaporized.

Typically Watson characterization factor varies between 10.5 and 13 for various crude streams. A highly paraffinic crude typically possesses a K factor of 13. On the other hand, a highly naphthenic crude possesses a K factor of 10.5. Therefore, Watson characterization factor can be used to judge upon the quality of the crude oil in terms of the dominance of the paraffinic or naphthenic components.

Sulfur Content

Since crude oil is obtained from petroleum reservoirs, sulphur is present in the crude oil. Usually, crude oil has both organic and inorganic sulphur in which the inorganic sulphur dominates the composition. Typically, crude oils with high sulphur content are termed as sour crude. On the other hand, crude oils with low sulphur content are termed as sweet crude. Typically, crude oil sulphur content consists of 0.5 – 5 wt % of sulphur. Crudes with sulphur content lower than 0.5 wt % are termed as sweet crudes. It is estimated that about 80 % of world crude oil reserves are sour.

The sulphur content in the crude oil is responsible for numerous hydrotreating operations in the refinery process. Strict and tighter legislations enforce the production of various consumer petroleum products with low quantities of sulphur (in the range of ppm). Presently, India is heading towards the generation of diesel with Euro III standards that indicates that the maximum sulphur content is about 500 ppm in the product. This indicates that large quantities of inorganic sulphur needs to be removed from the fuel. Typically, inorganic sulphur from various intermediate product streams is removed using hydrogen as hydrogen sulphide.

A typical refinery consists of good number of hydrotreaters to achieve the desired separation. The hydrotreaters in good number are required due to the fact that the processing conditions for various refinery intermediate process streams are significantly different and these streams cannot be blended together as well due to their diverse properties which were achieved using the crude distillation unit.

TBP/ASTM Distillation Curves

The most important characterization properties of the crude/intermediate/product streams are the TBP/ASTM distillation curves. Both these distillation curves are measured at 1 atm pressure. In both these cases, the boiling points of various volume fractions are being measured. However, the basic difference between TBP curve and ASTM distillation curve is that while TBP curve is measured using batch distillation apparatus consisting of no less than 100 trays and very high reflux ratio, the ASTM distillation is measured in a single stage apparatus without any reflux. Therefore, the ASTM does not indicate a good separation of various components and indicates the operation of the laboratory setup far away from the equilibrium.

Viscosity

Viscosity is a measure of the flow properties of the refinery stream. Typically in the refining industry, viscosity is measured in terms of centistokes (termed as cst) or saybolt seconds or redwood seconds. Usually, the viscosity measurements are carried out at 100° F and 210° F. Viscosity is a very important property for the heavy products obtained from the crude oil. The viscosity acts as an important characterization property in the blending units associated to heavy products such as bunker fuel. Typically, viscosity of these products is specified to be within a specified range and this is achieved by adjusting the viscosities of the streams entering the blending unit.

Flash and Fire Point

Flash and fire point are important properties that are relevant to the safety and transmission of refinery products. Flash point is the temperature above which the product flashes forming a mixture capable of inducing ignition with air. Fire point is the temperature well above the flash point where the product could catch fire. These two important properties are always taken care in the day to day operation of a refinery.

Pour Point

When a petroleum product is cooled, first a cloudy appearance of the product occurs at a certain temperature. This temperature is termed as the cloud point. Upon further cooling, the product will ceases to flow at a temperature. This temperature is termed as the pour point. Both pour and cloud points are important properties of the product streams as far as heavier products are concerned. For heavier products, they are specified in a desired range and this is achieved by blending appropriate amounts of lighter intermediate products.

Octane Number

Though irrelevant to the crude oil stream, the octane number is an important property for many intermediate streams that undergo blending later on to produce automotive gasoline, diesel etc.Typically gasoline tends to knock the engines. The knocking tendency of the gasoline is defined in terms of the maximum compression ratio of the engine at which the knock occurs. Therefore, high quality gasoline will tend to knock at higher compression ratios and vice versa. However, for comparative purpose, still one needs to have a pure component whose compression ratio is known for knocking. Iso-octane is eventually considered as the barometer for octane number comparison. While iso-octane was given an octane number of 100, n-heptane is given a scale of 0. Therefore, the octane number of a fuel is equivalent to a mixture of a iso-octane and n-heptane that provides the same compression ratio in a fuel engine. Thus an octane number of 80 indicates that the fuel is equivalent to the performance characteristics in a fuel engine fed with 80 vol % of isooctane and 20 % of n-heptane.

Octane numbers are very relevant in the reforming, isomerisation and alkylation processes of the refining industry. These processes enable the successful reactive transformations to yield long side chain paraffins and aromatics that possess higher octane numbers than the feed constituents which do not consist of higher quantities of constituents possessing straight chain paraffins and non-aromatics (naphthenes).

Crude Chemistry

Fundamentally, crude oil consists of 84 – 87 wt % carbon, 11 – 14 % hydrogen, 0 – 3 wt % sulphur, 0 – 2 wt % oxygen, 0 – 0.6 wt % nitrogen and metals ranging from 0 – 100 ppm. Understanding thoroughly the fundamentals of crude chemistry is very important in various refining processes. The existence of compounds with various functional groups and their dominance or reduction in various refinery products is what is essentially targeted in various chemical and physical processes in the refinery.

Based on chemical analysis and existence of various functional groups, refinery crude can be broadly categorized into about 9 categories summarized as

3.5.1 Paraffins: Paraffins refer to alkanes such as methane, ethane, propane, n and iso butane, n and iso pentane. These compounds are primarily obtained as a gas fraction from the crude distillation unit.

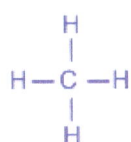 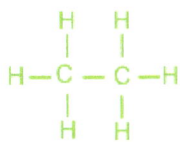 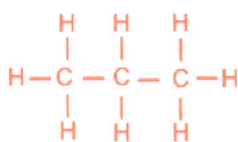 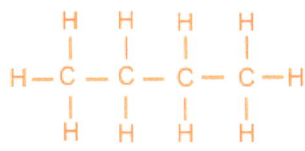

Methane(CH_4)   Ethane(C_2H_6)   Propene(C_3H_8)   Normal Butane(nC_4H_{10})

Normal Pentene (C_5H_{12})

Olefins: Alkenes such as ethylene, propylene and butylenes are highly chemically reactive. They are not found in mentionable quantities in crude oil but are encountered in some refinery processes such as alkylation.

Ethylene(C_2H_4)   Propylene(C_3H_6)   Butylene(C_4H_8)

Naphthenes: Naphthenes or cycloalkanes such as cyclopropane, methyl cyclohexane are also present in the crude oil. These compounds are not aromatic and hence do not contribute much to the octane number. Therefore, in the reforming reaction, these compounds are targeted to generate aromatics which have higher octane numbers than the naphthenes.

Cyclopropane(C_3H_6)   Cyclobutane(C_4H_8)   Cyclopentane(C_5H_{10})

Cyclohexane(C_6H_{12})   Methyl Cyclohexane(C_7H_{14})

Aromatics: Aromatics such as benzene, toluene o/m/p-xylene are also available in the crude oil. These contribute towards higher octane number products and the target is to maximize their quantity in a refinery process.

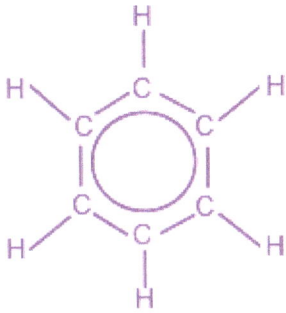

Benzene(C_6H_6)

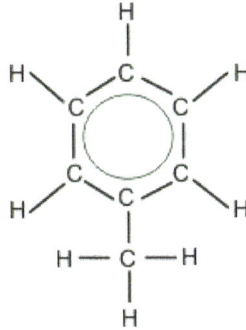

Tolune(C_7H_8)

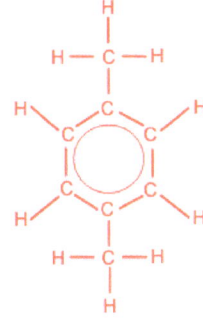

Para-X ylene(C_8H_{10})

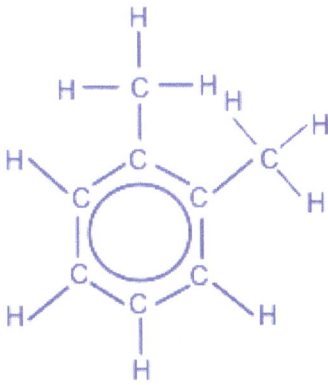

Ortho-Xylene(C_8H_{10})

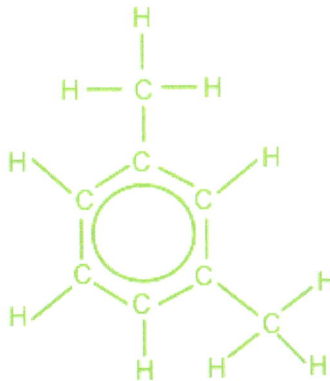

Meta-X ylene(C_8H_{10})

Napthalenes: Polynuclear aromatics such as naphthalenes consist of two or three or more aromatic rings. Their molecular weight is usually between 150 – 500.

Napthalenes

Organic sulphur compounds: Not all compounds in the crude are hydrocarbons consisting of hydrogen and carbon only. Organic sulphur compounds such as thiophene, pyridine also exist in the crude oil. The basic difficulty of these organic sulphur compounds is the additional hydrogen requirements in the hydrotreaters to meet the euro III standards. Therefore, the operating conditions of the hydrotreaters is significantly intense when compared to those that do not target the reduction in the concentration of these organic sulphur compounds. Therefore, ever growing environmental legislations indicate technology and process development/improvement on the processing of organic sulphur compounds.

Oxygen containing compounds: These compounds do not exist 2 % by weight in the crude oil. Typical examples are acetic and benzoic acids. These compounds cause corrosion and therefore needs to be effectively handled.

Resins: Resins are polynuclear aromatic structures supported with side chains of paraffins and small ring aromatics. Their molecular weights vary between 500 – 1500. These compounds also contain sulphur, nitrogen, oxygen, vanadium and nickel.

Asphaltenes: Asphaltenes are polynuclear aromatic structures consisting of 20 or more aromatic rings along with paraffinic and naphthenic chains. A crude with high quantities of resins and asphaltenes (heavy crude) is usually targeted for coke production.

In this section, a brief overview of various refinery processes is presented along with a simple sketch of the process block diagram of a modern refinery. The sketch of the modern refinery indicates the underlying complexity and the sketch is required to have a good understanding of the primary processing operations in various sub-processes and units.

Refinery Flow Sheet

We now present a typical refinery flowsheet for the refining of middle eastern crude oil. There are about 22 units in the flowsheet which themselves are complex enough to be regarded as process flow sheets. Further, all streams are numbered to summarize their significance in various processing steps encountered in various units. However, for the convenience of our understanding, we present them as units or blocks which enable either distillation in sequence or reactive transformation followed by distillation sequences to achieve the desired products.

The 22 units presented in the refinery process diagram are categorized as:

1. Crude distillation unit (CDU)

2. Vacuum distillation unit (VDU)

3. Thermal cracker

4. Hydrotreaters

5. Fluidized catalytic cracker

6. Separators

7. Naphtha splitter

8. Reformer

9. Alkylation and isomerisation

10. Gas treating

11. Blending pools

12. Stream splitters

A brief account of the above process units along with their functional role is presented next with simple conceptual block diagrams representing the flows in and out of each unit.

a. Crude distillation unit

The unit comprising of an atmospheric distillation column, side strippers, heat exchanger network, feed de-salter and furnace as main process technologies enables the separation of the crude into its various products. Usually, five products are generated from the CDU namely gas + naphtha, kerosene, light gas oil, heavy gas oil and atmospheric residue. In some refinery configurations, terminologies such as gasoline, jet fuel and diesel are used to represent the CDU products which are usually fractions emanating as portions of naphtha, kerosene and gas oil. Amongst the crude distillation products, naphtha, kerosene have higher product values than gas oil and residue. On the other hand, modern refineries tend to produce lighter components from the heavy products. Therefore, reactive transformations (chemical processes) are inevitable to convert the heavy intermediate refinery streams into lighter streams.

Operating Conditions : The temperature at the entrance of the furnace where the crude enters is 200 – 280°C. It is then further heated to about 330 – 370°C inside the furnace. The pressure maintained is about 1 barg.

b. Vacuum distillation unit (VDU)

The atmospheric residue when processed at lower pressures does not allow decomposition of the atmospheric residue and therefore yields LVGO, HVGO and vacuum residue. The LVGO and HVGO are eventually subjected to cracking to yield even lighter products. The VDU consists of a main vacuum distillation column supported with side strippers to produce the desired products. Therefore, VDU is also a physical process to obtain the desired products.

Operating Conditions : The pressure maintained is about 25 – 40 mm Hg. The temperature is kept at around 380 – 420°C.

c. Thermal cracker

Thermal cracker involves a chemical cracking process followed by the separation using physical principles (boiling point differences) to yield the desired products. Thermal cracking yields naphtha + gas, gasoil and thermal cracked residue. In some petroleum refinery configurations, thermal cracking process is replaced with delayed coking process to yield coke as one of the petroleum refinery products.

Operating Conditions : The temperature should be kept at around 450 – 500°C for the larger hydrocarbons to become unstable and break spontaneously. A 2-3 bar pressure must be maintained.

d. Hydrotreaters

For many refinery crudes such as Arabic and Kuwait crudes, sulfur content in the crude is significantly high. Therefore, the products produced from CDU and VDU consist of significant amount of sulfur.Henceforth, for different products generated from CDU and VDU, sulfur removal is accomplished to remove sulfur as H_2S using Hydrogen.The H_2 required for the hydrotreaters is obtained from the reformer unit where heavy naphtha is subjected to reforming to yield high octane number reforme product and reformer H_2 gas. In due course of process, H_2S is produced. Therefore, in industry, to accomplish sulfur removal from various CDU and VDU products, various hydrotreaters are used. In due course of hydrotreating in some hydrotreaters products lighter than the feed are produced. For instance, in the LVGO/HVGO hydrotreater, desulfurization of LVGO & HVGO (diesel) occurs in two blocked operations and desulfurized naphtha fraction is produced along with thedesulfurized gas oil main product. Similarly, for LGO hydrotreating case, along with diesel main product, naphtha and gas to C5 fraction are obtained as other products. Only for kerosene hydrotreater, no lighter product is produced in the hydrotreating operation. It is further interesting to note that naphtha hydrotreater is fed with both light and heavy naphtha as feed which is desulfurized with the reformer off gas. In this process, light ends from the reformer gas are stripped to enhance the purity of hydrogen to about 92 %. Conceptually, hydrotreating is regarded as a combination of chemical and physical processes.

Operating Conditions: The operating conditions of a hydrotreater varies with the type of feed. For Naphtha feed, the temperature may be kept at around 280-425°C and the pressure be maintained at 200 – 800 psig.

e. Fluidized catalytic cracker

The unit is one of the most important units of the modern refinery. The unit enables the successful transformation of desulfurized HVGO to lighter products such as unsaturated light ends, light cracked naphtha, heavy cracked naphtha, cycle oil and slurry. Thereby, the unit is useful to generate more lighter products from a heavier lower value intermediate product stream. Conceptually, the unit can be regarded as a combination of chemical and physical processes.

f. Separators

The gas fractions from various units need consolidated separation and require stage wise separation of the gas fraction. For instance, C4 separator separates the desulfurized naphtha from all saturated light ends greater than or equal to C4s in composition. On the other hand, C3 separator separates butanes (both iso and n-butanes) from the gas fraction. The butanes thus produced are of necessity in isomerization reactions, LPG and gasoline product generation. Similarly, the C2 separator separates the saturated C3 fraction that is required for LPG product generation and generates the fuel gas + H_2S product as well. All these units are conceptually regarded as physical processes.

Operating Conditions: Most oil and gas separators operate in the pressure range of 20 − 1500 psi.

g. Naphtha splitter

The naphtha splitter unit consisting of a series of distillation columns enables the successful separation of light naphtha and heavy naphtha from the consolidated naphtha stream obtained from several sub-units of the refinery complex. The naphtha splitter is regarded as a physical process for modeling purposes.

Operating Conditions: The pressure is to be maintained between 1 kg/cm² to 4.5 kg/cm². The operating temperature range should be 167 − 250°C

h. Reformer

As shown in Figure, Heavy naphtha which does not have high octane number is subjected to reforming in the reformer unit to obtain reformate product (with high octane number), light ends and reformer gas (hydrogen). Thereby, the unit produces high octane number product that is essential to produce premium grade gasoline as one of the major refinery products. A reformer is regarded as a combination of chemical and physical processes.

Operating Conditions : The initial liquid feed should be pumped at a reaction pressure of 5 − 45 atm, and the preheated feed mixture should be heated to a reaction temperature of 495 − 520°C.

i. Alkylation & Isomerization

The unsaturated light ends generated from the FCC process are stabilized by alkylation process using iC_4 generated from the C4 separator. The process yields alkylate product which has higher octane number than the feed streams. As isobutane generated from the separator is enough to meet the demand in the alkylation unit, isomerization reaction is carried out in the isomerization unit to yield the desired make up iC_4.

j. Gas treating

The otherwise not useful fuel gas and H_2S stream generated from the C2 separator has significant amount of sulfur. In the gas treating process, H_2S is successfully transformed into sulfur along with the generation of fuel gas. Eventually, in many refineries, some fuel gas is used for furnace applications within the refinery along with fuel oil (another refinery product generated from the fuel oil pool) in the furnace associated to the CDU.

Operating Conditions: Gas treaters may operate at temperatures ranging from 150 psig (low pressure units) to 3000 psig (high pressure units).

k. Blending pools

All refineries need to meet tight product specifications in the form of ASTM temperatures, viscosities, octane numbers, flash point and pour point. To achieve desired products with minimum specifications of these important parameters, blending is carried out. There are four blending pools in a typical refinery. While the LPG pool allows blending of saturated C3s and C4s to generate C3 LPG and C4 LPG, which do not allow much blending of the feed streams with one another. The most important blending pool in the refinery complex is the gasoline pool where in both premium and regular gasoline products are prepared by blending appropriate amounts of n-butane, reformate, light naphtha, alkylate and light cracked naphtha. These two products are by far the most profit making products of the modern refinery and henceforth emphasis is there to maximize their total products while meeting the product specifications. The gasoil pool produces automotive diesel and heating oil from kerosene (from CDU), LGO, LVGO and slurry. In the fuel oil pool, haring diesel, heavy fuel oil and bunker oil are produced from LVGO, slurry and cracked residue.

l. Stream splitters

To facilitate stream splitting, various stream splitters are used in the refinery configuration. A kerosene splitter is used to split kerosene between the kerosene product and the stream that is sent to the gas oil pool. Similarly, butane splitter splits the n-butane stream into butanes entering LPG pool, gasoline pool and isomerization unit. Unlike naphtha splitter, these two splitters facilitate stream distribution and do not have any separation processes built within them.

With these conceptual diagrams to represent the refinery, the refinery block diagram with the complicated interaction of streams is presented in Figure. A concise summary of stream description is presented in Table.

Atmospheric Distillation of Crude Oil

Distillation of crude oil is typically performed under atmospheric pressure and under a vacuum. Low boiling fractions usually vaporize below 400 °C at atmospheric pressure without cracking the hydrocarbon compounds. Therefore, all the low boiling fractions of crude oil are separated by atmospheric distillation. A crude distillation unit (CDU) consists of pre-flash distillation column. The petroleum products obtained from the distillation process are light, medium, and heavy naphtha, kerosene, diesel, and oil residue.

Atmospheric Crude Distillation Unit

Crude oil obtained from the desalter at temperature of 250 °C–260 °C is further heated by a tube-still heater to a temperature of 350 °C–360 °C. The hot crude oil is then passed into a distillation column that allows the separation of the crude oil into different fractions depending on the difference in volatility. The pressure at the top is maintained at 1.2–1.5 atm so that the distillation can be carried out at close to atmospheric pressure, and therefore it is known as atmospheric distillation column.

The vapors from the top of the column are a mixture of hydrocarbon gases and naphtha, at a temperature of 120 °C–130 °C. The vapor stream associated with steam used at bottom of the column is condensed by the water cooler and the liquid collected in a vessel is known as reflux drum which is present at the top of the column. Some part of the liquid is returned to the top plate of the column as overhead reflux, and the remaining liquid is sent to a stabilizer column which separates gases from liquid naphtha. A few plates below the top plate, the kerosene is obtained as product at a temperature of 190 °C–200 °C. Part of this fraction is returned to the column after it is cooled by a heat exchanger. This cooled liquid is known as circulating reflux, and it is important to control the heat load in the column. The remaining crude oil is passed through a side stripper which uses steam to separate kerosene. The kerosene obtained is cooled and collected in a storage tank as raw kerosene, known as straight run kerosene that boils at a range of 140 °C–270 °C. A few plates below the kerosene draw plate, the diesel fraction is obtained at a temperature of 280 °C–300 °C. The diesel fraction is then cooled and stored. The top product from the atmospheric distillation column is a mixture of hydrocarbon gases, e.g., methane, ethane, propane, butane, and naphtha vapors. Residual oil present at the bottom of the column is known as reduced crude oil (RCO). The temperature of the stream at the bottom is 340 °C–350 °C, which is below the cracking temperature of oil.

Simulation helps in crude oil characterization so that thermodynamic and transport properties can be predicted. Dynamic models help in examining the relationships that could not be found by experimental methods (Ellner & Guckenheimer, 2006). By using modeling and simulation software, 80% of the time can be saved rather than constructing an actual working model. Also it saves cost. Moreover, a model can provide more accurate study of the real system.

Cracking (Chemistry)

In petroleum geology and chemistry, cracking is the process whereby complex organic molecules such as kerogens or long chain hydrocarbons are broken down into simpler molecules such as light hydrocarbons, by the breaking of carbon-carbon bonds in the precursors. The rate of cracking and the end products are strongly dependent on the temperature and presence of catalysts. Cracking is the breakdown of a large alkane into smaller, more useful alkanes and alkenes. Simply put, hydrocarbon cracking is the process of breaking a long-chain of hydrocarbons into short ones. This process might require high temperatures and high pressure.

More loosely, outside the field of petroleum chemistry, the term "cracking" is used to describe any type of splitting of molecules under the influence of heat, catalysts and solvents, such as in processes of destructive distillation or pyrolysis.

Fluid catalytic cracking produces a high yield of petrol and LPG, while hydrocracking is a major source of jet fuel, Diesel fuel, naphtha, and again yields LPG.

Refinery using the Shukhov cracking process, Baku, Soviet Union, 1934

History and Patents

Among several variants of thermal cracking methods (variously known as the "Shukhov cracking process", "Burton cracking process", "Burton-Humphreys cracking process", and "Dubbs cracking process") Vladimir Shukhov, a Russian engineer, invented and patented the first in 1891 (Russian Empire, patent no. 12926, November 27, 1891). One installation was used to a limited extent in Russia, but development was not followed up. In the first decade of the 20th century the American engineers William Merriam Burton and Robert E. Humphreys independently developed and patented a similar process as U.S. patent 1,049,667 on June 8, 1908. Among its advantages was the fact that both the condenser and the boiler were continuously kept under pressure.

In its earlier versions however, it was a batch process, rather than continuous, and many patents were to follow in the USA and Europe, though not all were practical. In 1924, a delegation from the American Sinclair Oil Corporation visited Shukhov. Sinclair Oil apparently wished to suggest that the patent of Burton and Humphreys, in use by Standard Oil, was derived from Shukhov's patent for oil cracking, as described in the Russian patent. If that could be established, it could strengthen the hand of rival American companies wishing to invalidate the Burton-Humphreys patent. In the event Shukhov satisfied the Americans that in principle Burton's method closely resembled his 1891 patents, though his own interest in the matter was primarily to establish that "the Russian oil industry could easily build a cracking apparatus according to any of the described systems without being accused by the Americans of borrowing for free".

At that time, just a few years after the Russian Revolution, Russia was desperate to develop industry and earn foreign exchange, so their oil industry eventually did obtain much of their technology from foreign companies, largely American. At about that time however, fluid catalytic cracking was being explored and developed and soon replaced most of the purely thermal cracking processes in the fossil fuel processing industry. The replacement was however not complete; many types of cracking, including pure thermal cracking, still are in use, depending on the nature of the feedstock and the products required to satisfy market demands. Thermal cracking remains important however, for example in producing naphtha, gas oil, and coke, and more sophisticated forms of thermal cracking have been developed for various purposes. These include visbreaking, steam cracking, and coking.

Chemistry

A large number of chemical reactions take place during the cracking process, most of them based on free radicals. Computer simulations aimed at modeling what takes place during steam cracking have included hundreds or even thousands of reactions in their models. The main reactions that take place include:

Initiation

In these reactions a single molecule breaks apart into two free radicals. Only a small fraction of the feed molecules actually undergo initiation, but these reactions are necessary to produce the free radicals that drive the rest of the reactions. In steam cracking, initiation usually involves breaking a chemical bond between two carbon atoms, rather than the bond between a carbon and a hydrogen atom.

$$CH_3CH_3 \rightarrow 2\ CH_3\bullet$$

Hydrogen Abstraction

In these reactions a free radical removes a hydrogen atom from another molecule, turning the second molecule into a free radical.

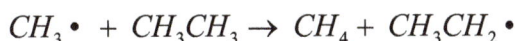

$$CH_3\bullet\ +\ CH_3CH_3 \rightarrow CH_4 + CH_3CH_2\bullet$$

Radical Decomposition

In these reactions a free radical breaks apart into two molecules, one an alkene, the other a free radical. This is the process that results in alkene products.

$$CH_3CH_2 \bullet \rightarrow CH_2 = CH_2 + H \bullet$$

Radical Addition

In these reactions, the reverse of radical decomposition reactions, a radical reacts with an alkene to form a single, larger free radical. These processes are involved in forming the aromatic products that result when heavier feedstocks are used.

$$CH_3CH_2 \bullet + CH_2 = CH_2 \rightarrow CH_3CH_2CH_2CH_2 \bullet$$

Termination

In these reactions two free radicals react with each other to produce products that are not free radicals. Two common forms of termination are *recombination*, where the two radicals combine to form one larger molecule, and *disproportionation*, where one radical transfers a hydrogen atom to the other, giving an alkene and an alkane.

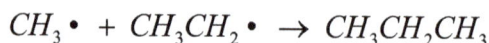

$$CH_3 \bullet + CH_3CH_2 \bullet \rightarrow CH_3CH_2CH_3$$

$$CH_3CH_2 \bullet + CH_3CH_2 \bullet \rightarrow CH_2 {=} CH_2 + CH_3CH_3$$

Example: Cracking Butane

There are three places where a butane molecule ($CH_3 - CH_2 - CH_2 - CH_3$) might be split. Each has a distinct likelihood:

- 48%: break at the $CH_3 - CH_2$ bond.

$$CH_3 * \ / \ * CH_2 - CH_2 - CH_3$$

    Ultimately this produces an alkane and an alkene: $CH_4 + CH_2 = CH - CH_3$

- 38%: break at a $CH_2 - CH_2$ bond.

$$CH_3 - CH_2 * \ / \ * CH_2 - CH_3$$

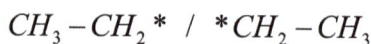

    Ultimately this produces an alkane and an alkene of different types:
$$CH_3 - CH_3 + CH_2 = CH_2$$

- 14%: break at a terminal $C - H$ bond

$$H / CH_2 - CH_2 - CH_2 - CH_3$$

Ultimately this produces an alkene and hydrogen gas:

$$CH_2 = CH - CH_2 - CH_3 + H_2$$

Cracking Methodologies

Thermal Methods

Thermal cracking was the first category of hydrocarbon cracking to be developed. Thermal cracking is an example of a reaction whose energetics are dominated by entropy ($\Delta S°$) rather than by enthalpy ($\Delta H°$) in the Gibbs Free Energy equation $\Delta G°=\Delta H°-T\Delta S°$. Although the bond dissociation energy D for a carbon-carbon single bond is relatively high (about 375 kJ/mol) and cracking is highly endothermic, the large positive entropy change resulting from the fragmentation of one large molecule into several smaller pieces, together with the extremely high temperature, makes $T\Delta S°$ term larger than the $\Delta H°$ term, thereby favoring the cracking reaction.

Thermal Cracking

Modern high-pressure thermal cracking operates at absolute pressures of about 7,000 kPa. An overall process of disproportionation can be observed, where "light", hydrogen-rich products are formed at the expense of heavier molecules which condense and are depleted of hydrogen. The actual reaction is known as homolytic fission and produces alkenes, which are the basis for the economically important production of polymers.

Thermal cracking is currently used to "upgrade" very heavy fractions or to produce light fractions or distillates, burner fuel and/or petroleum coke. Two extremes of the thermal cracking in terms of product range are represented by the high-temperature process called "steam cracking" or pyrolysis (ca. 750 °C to 900 °C or higher) which produces valuable ethylene and other feedstocks for the petrochemical industry, and the milder-temperature delayed coking (ca. 500 °C) which can produce, under the right conditions, valuable needle coke, a highly crystalline petroleum coke used in the production of electrodes for the steel and aluminium industries.

William Merriam Burton developed one of the earliest thermal cracking processes in 1912 which operated at 700–750 °F (371–399 °C) and an absolute pressure of 90 psi (620 kPa) and was known as the *Burton process*. Shortly thereafter, in 1921, C.P. Dubbs, an employee of the Universal Oil Products Company, developed a somewhat more advanced thermal cracking process which operated at 750–860 °F (399–460 °C) and was known as the *Dubbs process*. The Dubbs process was used extensively by many refineries until the early 1940s when catalytic cracking came into use.

Steam Cracking

Steam cracking is a petrochemical process in which saturated hydrocarbons are broken down into smaller, often unsaturated, hydrocarbons. It is the principal industrial method for producing the lighter alkenes (or commonly olefins), including ethene (or ethylene) and propene (or propylene). Steam cracker units are facilities in which a feedstock such as naphtha, liquefied petroleum gas (LPG), ethane, propane or butane is thermally cracked through the use of steam in a bank of pyrolysis furnaces to produce lighter hydrocarbons. The products obtained depend on the composition of the feed, the hydrocarbon-to-steam ratio, and on the cracking temperature and furnace residence time.

In steam cracking, a gaseous or liquid hydrocarbon feed like naphtha, LPG or ethane is diluted with steam and briefly heated in a furnace without the presence of oxygen. Typically, the reaction temperature is very high, at around 850 °C, but the reaction is only allowed to take place very briefly. In modern cracking furnaces, the residence time is reduced to milliseconds to improve yield, resulting in gas velocities up to the speed of sound. After the cracking temperature has been reached, the gas is quickly quenched to stop the reaction in a transfer line heat exchanger or inside a quenching header using quench oil.

The products produced in the reaction depend on the composition of the feed, the hydrocarbon to steam ratio and on the cracking temperature and furnace residence time. Light hydrocarbon feeds such as ethane, LPGs or light naphtha give product streams rich in the lighter alkenes, including ethylene, propylene, and butadiene. Heavier hydrocarbon (full range and heavy naphthas as well as other refinery products) feeds give some of these, but also give products rich in aromatic hydrocarbons and hydrocarbons suitable for inclusion in gasoline or fuel oil.

A higher cracking temperature (also referred to as severity) favors the production of ethene and benzene, whereas lower severity produces higher amounts of propene, C4-hydrocarbons and liquid products. The process also results in the slow deposition of coke, a form of carbon, on the reactor walls. This degrades the efficiency of the reactor, so reaction conditions are designed to minimize this. Nonetheless, a steam cracking furnace can usually only run for a few months at a time between de-cokings. Decokes require the furnace to be isolated from the process and then a flow of steam or a steam/air mixture is passed through the furnace coils. This converts the hard solid carbon layer to carbon monoxide and carbon dioxide. Once this reaction is complete, the furnace can be returned to service.

Catalytic Methods

The catalytic cracking process involves the presence of acid catalysts (usually solid acids such as silica-alumina and zeolites) which promote a heterolytic (asymmetric) breakage of bonds yielding pairs of ions of opposite charges, usually a carbocation and the very unstable hydride anion. Carbon-localized free radicals and cations are both highly unstable and undergo processes of chain rearrangement, C-C scission in position beta as in cracking, and intra- and intermolecular hydrogen transfer. In both types of processes, the corresponding reactive intermediates (radicals, ions) are permanently regenerated, and thus they proceed by a self-propagating chain mechanism. The chain of reactions is eventually terminated by radical or ion recombination.

Fluid Catalytic Cracking

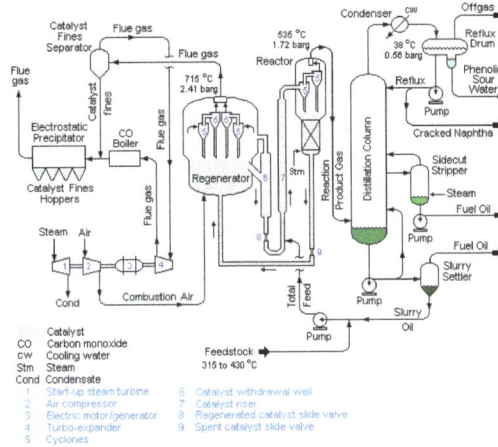

Schematic flow diagram of a fluid catalytic cracker

Fluid catalytic cracking is a commonly used process, and a modern oil refinery will typically include a *cat cracker*, particularly at refineries in the US, due to the high demand for gasoline. The process was first used around 1942 and employs a powdered catalyst. During WWII, the Allied Forces had plentiful supplies of the materials in contrast to the Axis Forces which suffered severe shortages of gasoline and artificial rubber. Initial process implementations were based on low activity alumina catalyst and a reactor where the catalyst particles were suspended in a rising flow of feed hydrocarbons in a fluidized bed.

Alumina-catalyzed cracking systems are still in use in high school and university laboratories in experiments concerning alkanes and alkenes. The catalyst is usually obtained by crushing pumice stones, which contain mainly aluminium oxide and silica into small, porous pieces. In the laboratory, aluminium oxide (or porous pot) must be heated.

In newer designs, cracking takes place using a very active zeolite-based catalyst in a short-contact time vertical or upward-sloped pipe called the "riser". Pre-heated feed is sprayed into the base of the riser via feed nozzles where it contacts extremely hot fluidized catalyst at 1,230 to 1,400 °F (666 to 760 °C). The hot catalyst vaporizes the feed and catalyzes the cracking reactions that break down the high-molecular weight oil into lighter components including LPG, gasoline, and diesel. The catalyst-hydrocarbon mixture flows upward through the riser for a few seconds, and then the mixture is separated via cyclones. The catalyst-free hydrocarbons are routed to a main fractionator for separation into fuel gas, LPG, gasoline, naphtha, light cycle oils used in diesel and jet fuel, and heavy fuel oil.

During the trip up the riser, the cracking catalyst is "spent" by reactions which deposit coke on the catalyst and greatly reduce activity and selectivity. The "spent" catalyst is disengaged from the cracked hydrocarbon vapors and sent to a stripper where it contacts steam to remove hydrocarbons remaining in the catalyst pores. The "spent" catalyst then flows into a fluidized-bed regenerator where air (or in some cases air plus oxygen) is used to burn off the coke to restore catalyst activity and also provide the necessary heat for the next reaction cycle, cracking being an endothermic reaction. The "regenerated" catalyst then flows to the base of the riser, repeating the cycle.

The gasoline produced in the FCC unit has an elevated octane rating but is less chemically stable compared to other gasoline components due to its olefinic profile. Olefins in gasoline are responsible for the formation of polymeric deposits in storage tanks, fuel ducts and injectors. The FCC LPG is an important source of C_3-C_4 olefins and isobutane that are essential feeds for the alkylation process and the production of polymers such as polypropylene.

Hydrocracking

Hydrocracking is a catalytic cracking process assisted by the presence of added hydrogen gas. Unlike a hydrotreater, where hydrogen is used to cleave C-S and C-N bonds, hydrocracking uses hydrogen to break C-C bonds (hydrotreatment is conducted prior to hydrocracking to protect the catalysts in a hydrocracking process).

The products of this process are saturated hydrocarbons; depending on the reaction conditions (temperature, pressure, catalyst activity) these products range from ethane, LPG to heavier hydrocarbons consisting mostly of isoparaffins. Hydrocracking is normally facilitated by a bifunctional catalyst that is capable of rearranging and breaking hydrocarbon chains as well as adding hydrogen to aromatics and olefins to produce naphthenes and alkanes.

The major products from hydrocracking are jet fuel and diesel, but low sulphur naphtha fractions and LPG are also produced. All these products have a very low content of sulfur and other contaminants.

It is very common in Europe and Asia because those regions have high demand for diesel and kerosene. In the US, fluid catalytic cracking is more common because the demand for gasoline is higher.

The hydrocracking process depends on the nature of the feedstock and the relative rates of the two competing reactions, hydrogenation and cracking. Heavy aromatic feedstock is converted into lighter products under a wide range of very high pressures (1,000-2,000 psi) and fairly high temperatures (750°-1,500 °F, 400-800 °C), in the presence of hydrogen and special catalysts.

The primary function of hydrogen is, thus:

1) preventing the formation of polycyclic aromatic compounds if feedstock has a high paraffinic content.

2) reduced tar formation

3) reducing impurities.

4) preventing buildup of coke on the catalyst.

5) converting sulfur and nitrogen compounds present in the feedstock to hydrogen sulfide and ammonia, and

6) achieving high cetane number fuel.

With the continuous depletion in world oil reserves and increasing demand of petroleum products, the refiners are forced to process more and heavier crude [Tondon et al., 2007. The cost advantage of heavy crudes over light crudes has incentivized many Indian Refineries to process heavier crude, therefore increasing the heavy residue produced at a time when fuel oil demand is declining [Haizmann et al., 2012]. In order to dovetail both the requirement for processing crude oil of deteriorating quality and enhancing distillates of improved quality, technological upgradation have been carried out at refineries which takes care of processing heavy crudes as well as maximizing value added products and stringent product quality requirements. Some of the Residue Upgradation Technologies in Indian Refineries is given in Table.

Table : Residue Upgradation Technologies in Refineries

Delayed Coking and Visbreaking	Technology for the bottom of the barrel upgradation; means of disposing of low value resids by converting part of the resids to more valuable liquid and gas products.
Uniflex Technology:	Technology for processing low quality residue by thermal cracking to produce high quality distillate products.
Fluidized Catalytic Cracking (FCC) and Residual Fluidized Catalytic Cracking (RFCC)	A technology introduced to contain generation of black oil from refinery and to increase the production of high value products like LPG, MS and Diesel.
Hydrocracking	Processes for Residue up gradation FCC process, delayed coking process &visbreaking
Deep Catalytic Cracking (DCC) and IOCL's INDMAX	For selectively cracking feed stocks to light olefins.

Cracking

Cracking of heavy residue is most commonly used method for upgradation of residues. This involves of decomposition of heavy residues by exposure to extreme temperatures in the presence or absence of catalysts.

THERMAL CRACKING: Cracking at elevated temperatures in the absence of catalyst eg: Visbreaking, delayed coking, Fluid coking etc.

Catalytic Cracking: Cracking in presence of catalyst eg: FCC , Hydrocracking, DCC

Cracking Mechanism

Cracking takes Place by Free Radical Mechanism.

Initiation

$$C_6H_{14} \rightarrow C_2H_5{}^O + C_4H_9{}^O$$

Propagation

$$C_2H_5{}^O + C_6H_{14} \rightarrow C_2H_6 + C_6H_{13}{}^O$$

$$C_4H_9{}^O + C_6H_{14} \rightarrow C_4H_{10} + C_6H_{13}{}^O$$

$$C_4H_9{}^O \rightarrow C_3H_6 + CH_3{}^O$$

$$C_6H_{13}{}^O \rightarrow C_4H_8 + C_2H_5{}^O \ (MANY \ OTHER \ PRODUCTS)$$

TERMINATION

$$C_2H_5{}^O + CH_3{}^O \rightarrow C_3H_8$$

During the cracking operation, some coke is usually formed. Coke is the end product of polymerisation reaction in which two large olefin molecules combine to form an even larger olefinic molecule

$$C_{10}H_{21} - CH = CH_2 + CH_2 = CH - C_{10}H_{21} \rightarrow C_{10}H_{21} - CH = CH - CH_2 - CH_2 - C_{10}H_{21}$$

Thermal Cracking

Thermal cracking process for upgradation of heavy residue has been used since long and still it is playing an important role in the modern refinery through upgradation of heavy residue and improving the economics of the refinery through the production of lighter distillate and other valuable product like low value fuel gas and petroleum coke. Although petroleum coke was first made by North Western Pennsylvanian the 1860's using cracking, however, a real breakthrough in the thermal cracking process was with development of the first cracker by William Burton and first used in 1913.Heavy residues are a mixture molecules consisting of an oil phase and an asphaltene phase in physical equilibrium with each other in colloidal form.

Asphaltenes are high molecular weight, relatively high atomicity molecules containing high levels of metals. During thermal cracking, the long molecules thus depleting the oil phase in the residue.

Asphaltene cracking is the most difficult component to process and asphaltenes in the feed remain unaffected, additional asphaltenes may be formed by secondary polymerisation reactions.

At a certain condition asphaltenes is disturbed and asphaltenes precipitate. At this stage of conversion the product residue becomes unstable.

Under condition of thermal cracking, hydrocarbons, when heated, decompose into smaller hydrocarbon molecules. The UOP thermal cracking process is based on two coil design for selective cracking of topped or reduced crudes into valuable products >350°C. Table shows the various thermal cracking process and process conditions.

Higher Boiling Petroleum Stock → Lower Boiling Products

Free radical chain reaction:

Free radical + Hydrocarbon → Stable End Product

Thermal Treatment; Medium, High,Ultra High (Cracking with higher Temperature and with very short residence time)

Reactions

-Cracking of side chains free aromatic group

-Dehydrogenation of naphthenes to form aromatics.

-Condensation of aliphatic to form aromatics.

-Condensation of aromatic to form higheraromatics.

-Dimerisation or oligomerisation

Development of Cracking Processes:

Year	Process
1861	Thermal cracking
1910	Batch Thermal Cracking
1912	Burton Cracking
1914-22	Continuous Cracking Process

PROCESS VARIABLES: Feed stock properties, Cracking Temperature, Residence time, PressureThermal Cracking: Medium, High, Ultra High (Cracking with higher Temperature and with very short residence time)

Table : Various Thermal Cracking Process and Process Conditions

Process	Process conditions
Visbreaking	Mild thermal cracking (low severity)
	Mild (470-500°C) heating at 50-200 psig
	Improve the viscosity of fuel oil
	Low conversion (10%) to 430°F
	Residence time1-3 min
	Heated coil or drum
Delayed coking	Operates in semi batch mode
	Moderate (900-960°F) heating at 90 psig
	Soak drums (845-900°F) coke walls
	Coked until drum solid
	Coke (removed hydraulically) 20-40% on feed, Yield 430°F, 30%
Fluid Coking	Server (510-520°C) heating at 10 psig
	Oil contact refractory coke
	Bed fluidized with steam-even heating,
	Higher yield of light ends (<Cs),Less coke yield
Flexicoking	A continuous fluidised bed technology which converts heavy residue to lighter more valuable product. The process essentially eliminates the coke production. Temperature 525°C

Visbreaking

Visbreaking is essentially a mild thermal cracking operation at mild conditions where in long chain molecules in heavy feed stocks are broken into short molecules thereby leading to a viscosity reduction of feedstock. Now all the new visbreaker units are of the soaker type. Soaker drum utilizes a soaker drum in conjunction with a fired heater to achieve conversion [Sieli, 1998] Visbreaking is a non-catalytic thermal process. It reduces the viscosity and pour point of heavy petroleum fractions so that product can be sold as fuel oil. It gives 80 - 85% yield of fuel oil and balance recovered as light and middle distillates. The unit produces gas, naphtha, heavy naphtha, visbreaker gas oil, visbreaker fuel oil (a mixture of visbreaker gas oil and vsibreaker tar). A given conversion in visbreaker can be achieved by two ways:

• High temp., low residence time cracking: Coil Visbreaking.

- Low temp., high residence time cracking: Soaker visbreaking.

Reaction in Visbreaking

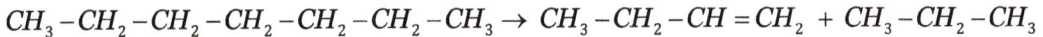

$$CH_3-CH_2-CH_2-CH_2-CH_2-CH_2-CH_3 \rightarrow CH_3-CH_2-CH=CH_2 + CH_3-CH_2-CH_3$$

Soaker Visbreaking Process

The furnace operators at a lower outlet temperature and a soaker drum is provided at the outlet of the furnace to give adequate residence time to obtain the desired conversion while producing a stable residue product, thereby increasing the heater run and reducing the frequency of unit shut down for heater decoking [Sieli,1998]. The products from soaker drum are quenched and distilled in the downstream fractionator. Process diagram for visbreaking with soaker is shown in Figure.

Objective: To lower the viscosity of heavy residues under relatively milder cracking condition than the conventional cracking processes.

Feed Atmospheric residues →To get gasoline and diesel oil

Vacuum residues →To reduce viscosity

Reaction:

-Splitting of C-C bond.

-Oligomerization and cyclisation to naphthenes of olefinic compounds.

-Condensation of the cyclic molecules to polyaromatics.

Side reactions: Foramation of H_2S, thiophenes, mercaptans, phenol

Products:

The cracked product contains gas, naphtha, gas oil and furnace oil, the composition of which will depend upon the type of feedstock processed. A typical yield pattern may be gas 1-2%, naphtha 2-3%, gas oil 5-7%, furnace oil 90-92%.

Visbreaking Conditions:

Inlet Temperature: 305-325°C (15-40 bar)

Exit: 480-500°C (2-10 bar)

With soaking 440-460°C (5-15 bar)

Feed: $90°C \rightarrow$ *pretreated with VB tar to* $335°C$

Visbreaking Furnace:

Convection zone → *top* → *to thermal efficiency*

Radiation zone → *bottom tubes.*

Avg. heat flow → $22 - 30 \, kw / m^2$

Variables:

Variables in visbreaker are feed rate, furnace transfer temperature, visbreaker tar quench to transfer line, fractionation pressure, fractionation top temperature, circulation, reflux flow, visbreaker tar quench to fractionator bottom, visbreaker tar quench to visbreaker tar stripper bottom, stabilizer temperature and pressure.

The purpose of visbreaking is to produce lower viscosity fuel oil.

Soaking Drum

Soaking drum is used to lengthen the feed residence time so that the furnace can operate at lower temperature. Soaker results in saving of energy because of the lower temperature with less coke tendency, larger gas oil yield

Advantages:

-15% reduction in fuel oil

-Larger running time between two decoking operations. coke deposit rate 3-4 times slower than in conventional units.

-Better selectivity towards gas and gasoline productivity.

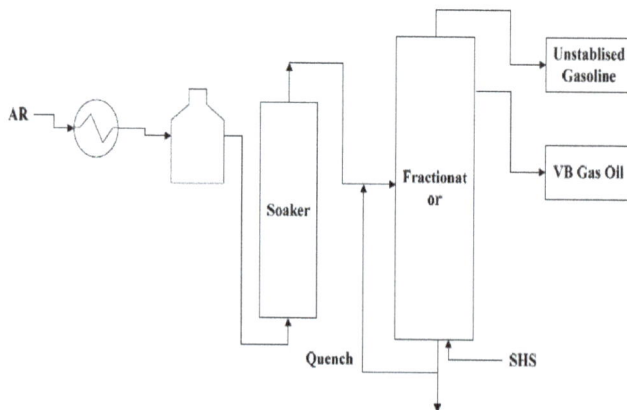

Visbreaking with Soaker

Coil Visbreaker: In coil visbreaking process the desired cracking is achieved in the furnace at high temperature and the products of cracking are quenched and distilled in a downstream fractionator. Advances in visbreaker coil heater design now allows for the isolation of one or more heater passes for decoking, eliminating the need of shut the entire visbreaker down for furnace decoking. Integration of the coil visbreaking process with vacuum units is also being considered in many areas of the world [Sieli,1998]

Coking

Coking is very severe form of thermal cracking and converts the heaviest low value residue to valuable distillates and petroleum coke. Relatively severe cracking operations to convert residual oil products and represents the complete conversion of petroleum residues to coke and lighter product. Recycle is used to further convert heavy distillate fractions to lighter products Mechanism of coke formation: The colloidal suspension of the asphaltenes and resin compounds is distorted, resulting in precipitation of highly cross linked structure of amorphous coke. The compounds are also subjected to cleavage of the aliphatic groups. Polymerisation andcondensation of the free aromatic radicals, grouping a large number of these compounds to such a degree that dense high grade coke is eventually formed.

The process involves thermal conversion of vacuum residue or other hydrocarbon residue resulting in fuel gas, LPG, naphtha, gas oil and coke and essentially a complete rejection of metals. Various types of coking processes are delayed coking, fluid coking and flexi coking.

Delayed Coking

Delayed coking process is used to crack heavy oils into more valuable light liquid products with less valuable gas and solid coke as byproducts. Although first delayed coking plant was built in1930, however delayed coking process has been evolving for 78 years, the past few years have seen changes in feed stock that has major impact on the design and operation of delayed coking [Catala, 2009].

Delayed coking consists of thermal cracking of heavy residue in empty drum where deposition of coke takes place. The product yield and quality depends on the typed feedstock processed. Typical delayed coking consists of a furnace to preheat the feed, coking drum where the fractionation of the product takes place.

The feed is first preheated in furnace where the desired cooking temperature is achieved and fed to the coking drums normally installed in pairs where the cracking reaction takes place and the coke is deposited in the bottom of the reactor. The coke drums overhead vapour flows to the fractionating column where they are separated into overhead streams containing wet gas LPG and naphtha and two side gas oil streams.

Recycled stream from the fractionating column combines with the fresh feed in the bottom of the column and is further preheated in coke heaters and flows to the coke drums. When a coke drum is filled, the heated streams from the coke heated are sent to the other drum. Process flow diagram for delay coking is shown in Figure.

The reaction involved in delayed coking is partial vaporization and partial cracking, cracking of two vapour phase in the coke drum and successive cracking used polymerization of liquid phaseresulting in formation of coke in the drum.

Feed: Vacuum residue, FCC residual, or cracked residue. Studies show that feedstock quality and severity on conversion impact the stability of visbreaker residue

Product: Gases, Naphtha, Fuel oil, Gas oil and Coke.

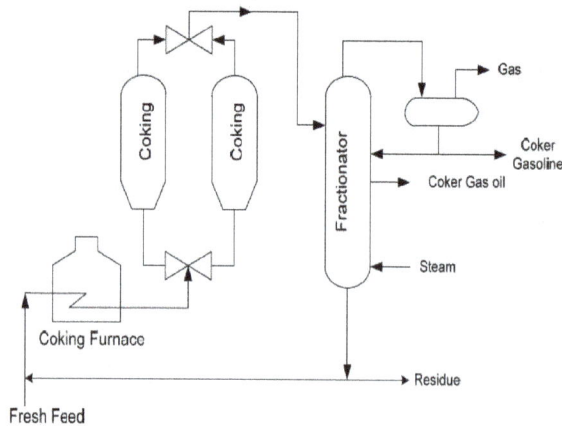

Process Flow Chart of Delayed Coaking

Operating Variables

Feed stocks variables: Chracterisation factor, degree of reduction, conardson carbon, sulphur content, Metallic constituents.Low feed stock characterisation factor and high carbon residue increase coke yield and quality of gas oil end point.

Engineering Variables: Batch, Semi continuous or continuous, Capacity and size factors, Coke removal equipment, Coke handling, Storage and Transportation

Operating Variables: Various operating variables in delayed coking are temperature, pressure, recycle ratio, transfer temperature, and coke chamber pressure.

Higher temperature: Results in more vaporization of the inlet material causing low coke yield. A furnace outlet temperature of 485-505 °C is maintained. High temperature results in hard coke while coke is soften when too low temperature is maintained.

Higher pressure: Results in an increase in coke and gas yields which is undesirable as basic objective is to improve the yield of distillation with less coke.

Higher recycle ratio: Results in higher coke and gas yields. Lower recycle ratio is always desirable for higher yield of liquid product.

Coke yield =1.6 CCR(conardson carbon residue)

Gas yield= 7.8+.144xCCR

Naphtha yield=11.29+0.343CCR

Gas yields =110-coke-gas yield-naphtha yield

Low feed stock characterization factor and high carbon residue increase coke yield and quality of gas oil end point.

Fluid Coking:Fluid coking is non- catalytic fluid bed process whereresidue is coked by spraying into a fluidized bed of hot, fine coke particles. Higher temperature with shorter contact time than delayed coking results in increased light and medium hydrocarbons with less cake generation. Shorter residence time can yield higher quantities of liquid less coke, but the product have lower value [Rana et al., 2007].

Flexi Coking:It is continuous process involves thermal cracking in a bed fluidized coke and gasification of the coke produced at 870 °C. This process contains an additional step of gasification(Temp: The gases leaving the gasifier is low calorific value fuel gas at 800-1500 kcal/m³ (4200 to 5000 kJ/std m3 and is burned in the furnace or power plants.It can be applied to wide variety of feed stocks.

UOP Uniflex™ Process : It is high conversion, commercially proven technology, that processes low quality residue streams, like vacuum residue, to make very high quality distillate products. The process utilizes thermal cracking to reduce molecular weight of the residue in the presence of hydrogen and a proven proprietary, nano sized catalyst to stabilize the cracked products and inhibit the formation of coke precursors. The main products from uniflex are naphtha and diesel with a yield of greater than 80 vol%.

Catalytic cracking process was developed in1920 by Eugene Houdry for upgradation of residue was commercialized latter in 1930. Houdry process was based on cyclic fixed bed configuration. There has been continuous upgradation in catalytic in catalytic cracking process from its incept of fixed bed technology to latter fluidized bed catalytic cracking (FCC).The feed stock for catalytic cracking is normally light gas oil from vacuum distillation column. Catalytic cracking cracks low value high molecular weight hydrocarbons to more value added products (low molecular weight) like gasoline, LPG Diesel along with very important petrochemical feedstock like propylene, C4 gases like isobutylene, Isobutane, butane and butane. Main Catalytic Cracking Reaction is given in Table.

Main Reactions Involved in Catalytic Cracking are

- Cracking

- Isomerisation

- Dehydrogenenation

- Hydrogen transfer

- Cyclization

- Condensation

- Alkylation and dealkylation

Major primary reactions taking place in catalytic cracking is given in Table

Paraffins → *Smaller paraffins + olefins*

Alkyl naphthene → *naphthene + olefin*

Alkyl aromatic → *aromatic + olefin*

Multiring naphthene → *alkylated naphthene with fewer rings*

Table : Main Catalytic Cracking Reaction

Paraffins	Cracking ⟶ Paraffins + Olefins
Olefins	Cracking ⟶ LPG Olefins
	Cyclization ⟶ Naphthenes

	Isomerization ⟶ Branched Olefins ⟶ H Transfer ⟶ Branched Paraffins
	H Transfer ⟶ Paraffins
	Cyclization ⟶ Coke
	Condensation ⟶ Coke
	Dehydrogenation ⟶ Coke
Naphthenes	Cracking ⟶ Olefins
	Dehydrogenation ⟶ Cyclo- Olefins ⟶ Dehydrogenation ⟶ Aromatics
	Isomerization ⟶ Naphthenes with different rings
Aromatics	Side chain cracking ⟶ Unsubstituted aromatics + Olefins
	Trans alkylation ⟶ Different alkyl aromatics
	Alkylation Dehydrogenation ⟶ Polyaromatics ⟶ Dehydrogenation ⟶ Coke Condensation
Hydrogen Transfer ⟶ Naphthene + Olefin ⟶ Aromatic + Paraffin	

Fluid Catalytic Cracking

Fluid catalytic cracking is now major secondary conversion process in Petroleum re-finery since 1942. there are more than 400 FCC units in world. The process provides around 50 percent of all transportation fuel and 35percent of total gasoline pool. Major land marks in the history of FCC has been:

- Introduction of zeolite catalyst during 1960 which has resulted in lower residence time

- Introduction of ultra stable Y-zeolite in mid 60's

- Switch over from bed cracking to riser cracking

- Introduction of large number of additives for boosting of gasoline octane/yield of light naphtha

- SOx control

- Nickel and vanadium passivation

FCC is a multi-component catalyst system with circulating fluid bed reactor system with reactor Regenerator system configuration. Figure shows details of FCC process and FCC reactor.

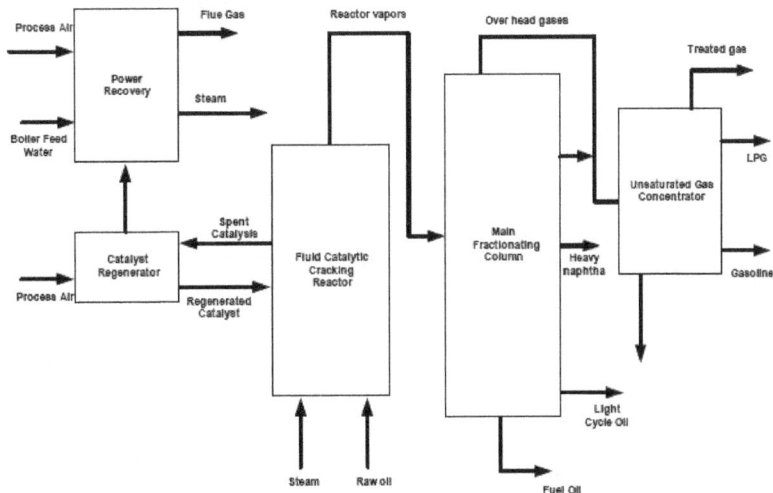

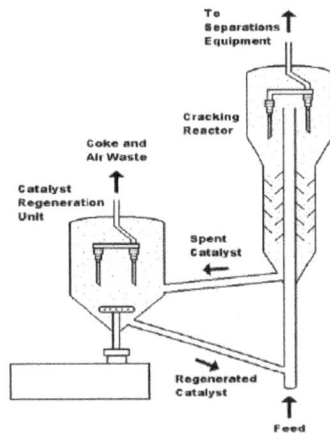

Fluid Catalytic Cracking Process and FCC Reactor

Feed Stock

Vacuum gas oil (VGO), Hydro-treated VGO, Hydro-cracker bottom, Coker gas oil (CGO), De- asphalted oil (DAO), Reduced crude oil (RCO), Vacuum residue (VR)

- Typical feedstock consists of Vacuum and Atmosphere gas oil but may include other heavy stream.

- Major contaminant in the feed includes carbon residue and metals.

- While FCC process feed containing up to 4% Conradson carbon MSCC can process all kinds of feed.

Process Steps

Three basic functions in the catalytic cracking process are:

Reaction - Feedstock reacts with catalyst and cracks into different hydrocarbons;

Regeneration - Catalyst is reactivated by burning off coke; and recerculated to reactor Fractionation - Cracked hydrocarbon stream is separated into various products like LPG and gasoline, like light cycle oil and heavy cycle oil are withdrawn as side stream

Reactor and Regenerator Section: Catalyst section consists mainly of the reactor and regenerator

- The feed to unit along with recycle streams is preheated to temperature of 365°C-370°C and enters the riser where it comes in contact with hot regenerated catalyst (at a temperature of about 640-660°C. Finely divided catalyst is maintained in an aerated or fluidized state by the oil vapors.

- The catalyst section contains the reactor and regenerator & catalyst re circulates between the two.

- Spent catalyst is regenerated to get rid of coke that collects on the catalyst during the process. Spent catalyst flows through the catalyst stripper to the regenerator, where most of the coke deposits burn off at the bottom where preheated air and spent catalyst are mixed. Fresh catalyst is added and worn-out catalyst removed to optimize the cracking process

Fractionation - Cracked hydrocarbon stream is separated into various products. LPG and gasile are removed overhead as vpour. Unconverted product like light cycle oil and heavy cycle oil are withdrawn as side stream. Overhead product is sent to stabilsation section where stablised gasoline is separated from light products from which LPG is recovered.

Typical Operating Parameter of FCC

Raw oil feed at heater inlet   :   114 cubic meter /h

Furnace outlet temperature   : 291^0C

Reactor feed temperature   : $371^\circ C$

Reactor Vapour temperature : $549^\circ C$

Product Obtained

- Light gas -H_2, C_1, and C_{2s}

- LPG C_{3s} and C_{4s} – includes light olefins

- Gasoline C_5+ high octane component for gasoline pool or light fuel

- Light cycle oil (LCO) blend component for diesel pool or light fuel

- Heavy cycle oil (HCO) Optional heavy cycle oil product for fuel oil or cutter stock

- Clarified oil (CLO) or decant oil: slurry for fuel oil

- Coke by-product consumed in the regenerator to provide the reactor heat demand

FCC Catalysts

Major breakthrough in the catalytic racking process was development zeolite catalysts which demonstrated vastly superior activity, gasoline selectivity, and stability characteristics compared to original amorphous silica alumina catalyst .

Year	1950	1970	1990
Zeolite content, wt%	0	10	Upto 40
Particle density, g/cc	0.9	1	1.4
Relative Attrition Index	20	5	1

Today's FCC catalysts Porous spray dried micro-spherical powder

- Particle size distribution of 20 -120 micron & particle density ~ 1400 kg/m³

- Comprising Y zeolite in many derivatives of varying properties

- Supplied under various grades of particle sizes & attrition resistance

- Continuing improvement metal tolerance, coke selectivity

New bread of catalyst are high metal tolerance with high matrix catalyst having better accessibility, regenerability and strippability .

Options for Clean Fuel

For upgrading FCC products into acceptable blending components following three steps are being used [Letzsch, 2005];

- Severe hydro processing of feed to FCC

- Treating each of the products in hydrotreater

- Combination of both upstream and downstream processing

Modified Catalytic Cracking Processes

Resid FCC (RFCC) Process

The RFCC process uses similar reactor technology as the FCC process and is targeted for residual feeds greater than 4 wt-% Conradson carbon. A two stage regenerator with catalyst cooling is typically used to control the higher coke production and resulting heat.

Deep Catalytic Cracking (DCC)

Milli Second Catalytic cracking (MSCC) Process:

Improvements in riser termination devices have led to significant decreases in post-riser residence time and post-riser cracking. The benefits of shorter catalyst-and oil contact time have been lower dry gas yields, lower delta coke on catalyst and more selective cracking to gasoline and light olefins.

- Due to improvement in reactor design there is lower regenerator temperature and higher catalyst recovery.

Petro FCC Process

The Petro FCC process targets the production of petrochemical feedstock rather than fuel products. This new process, which utilizes a uniquely designed FCC unit, can produce very high yields of light olefins and aromatics when coupled with aromatics complex. The catalyst section of the Petro FCC process uses a high-conversion, short-contact time reaction zone that operates at elevated reactor riser outlet temperatures

Indmax Technology- Residue to Olefin was developed by IOC R&D center and has been successfully commissioned in IOC Guwahati Unit [Bhatacharya, 2001]

Some of the special features of the technology are:

Operational features of Indmax technology

- Very high cat/oil ratio(15-25)

- Higher riser temperature (>550°C)

- Higher riser steam rate

- Relatively lower regeneration temperature.

Benefits

- LPG 35-65 wt%

- Propylene 17-25 wt% feed

- High octane gasoline (95+)

Multifunctional Proprietary Catalyst

- Higher propylene selectivity

- Superior metal tolerance

- Lower coke mate

Maximizing Propylene Output in FCC

New FCC processes are being operated to maximize the yield of propylene due to growing demand of propylene. Significant scope exists in the refinery in Asia region to enhance the production of propylene in Asia region [Ghosh, 2006]. Maximizing propylene yield from FCC is typically accomplished by combining a low rare earth catalyst system with severe reaction condition [Amano,T.,Wilcox,J. and Pouwels,C. " Process and catalysis factors to maximize propylene output" Petroleum Technology Quarterly 3, 2012, p.17]. some of the olefin maximizing technology are deep catalytic cracking (DCC) based on riser bed catalytic cracking, Propylene –Max technology by ABBS Lummus global, Maxofin Process by Mobil-M.W. Kellog.

Hydrocracking

The development of upgrading technology for heavier stocks having high sulfur, nitrogen and heavy metal (Ni, V) are becoming important. Hydrocracking is one of the most versatile processes for the conversion of low quality feed stocks into high quality products like gasoline, naphtha, kerosene, diesel, and hydrowax which can be used as petrochemical feed stock. Its importance is growing more as a refiners search for low investment option for producing clean fuel. New environmental legislations require increasing and expensive efforts to meet stringent product quality demands.

Hydrocracking processes uses a wide variety of feed stocks like naphtha, atmospheric gas oil, vacuum gas oils, coke oils, catalytically cracked light and heavy cycle oil, cracked residue, deasphalted oils and produces high quality product with excellent product quality with low sulfur contents. Comparison of catalytic cracking and hydro-cracking is given in Table.

The history of the hydrocracking process goes back to late 1920 when hydrocracking technology for coal conversion was developed in Germany. During World War II, two stage hydrocracking were applied in Germany, USA and Britain. However, real break-through in hydrocracking process was with the development of improved catalyst due to which processing at lower pressure. Hydrocracking can process wide variety of feed-stocks producing wide range of products.

Feed: Straight run gas oil, Vacuum gas oils, Cycle oils, Coker Gas oils, thermally cracked stocks, Solvent deasphalted residual oils, straight run naphtha, cracked naphtha.

Product: Liquefied petroleum gas (LPG), Motor gasoline, Reformer feeds, Aviation tur-bine fuel, Diesel fuels, heating oils, Solvent and thinners, Lube oil, FCC feed

Table: Comparison of Catalytic Cracking and Hydrocracking

Catalytic Cracking	Hydro-cracking
Carbon rejection	Hydrogen addition
Riser-regenerator-configuration	Downflow packed bed
LPG/gasoline	Kerosene/diesel
Product rich in unsaturated components	Few aromatics, low S- and N-content in product

Recent Development Hydrocracking

There has been continuous development in the hydrocracking technology both in pro-cess and catalyst. Some of the important development in hydrocracking has been mild hydrocracking and resid hydrocracking. Mild hydrcracking (MHC) is characterized by relatively low conversion (20-40%) as compared to convention hydrocracking which give 70-100% conversion of heavy distillate at high pressure. MHC route produces low sulphur (10 ppm sulphur as desired by future diesel specification) diesel. New mild hy-drocracking route produces 10 ppm sulphur diesel which is produced by hydro-crack-ing under mild pressure. MHC allows increasing diesel production through VGO hy-dro-conversion.

The yield of middle distillates obtained from hydro cracker is much more than that obtained from other processes also, hydro cracker does not yield coke or pitches as by product. The increased demand for environmentally acceptable products forced the refiners to accept stringent specifications for gasoline & diesel necessitating the use of hydrocracking technology to limit sulphur & aromatic in petroleum products. No post treatment is required for the hydro cracker products.

Hydro-treatment & Hydrocracking Catalyst

Hydrocracking processes involved two types of catalyst:

- Hydro pretreatment catalyst
- Hydrocracking catalyst

Hydrotreating (Pretreat) Catalyst

The main objective of pretreat catalyst is to remove organic nitrogen from the hydro cracker feed allowing

(i) Better performance of second stage hydrocracking catalyst, and

(ii) The initiation of the sequence of hydrocracking reactions by saturation of aromatic compounds

Pretreat catalyst must have adequate activity to achieve above objectives within the operating limits of the hydrogen partial pressure, temperature and LHSV.

Hydrocracking Catalyst

Hydrocracking catalyst is a bi-functional catalyst and has a cracking function and hydrogenation- dehydrogenation function. The former is provided by an acidic support whereas the latter is imparted by metals. Acid sites (Crystalline zeolite, amorphous silica alumina, mixture of crystalline zeolite and amorphous oxides) provide cracking activity. Metals [Noble metal (Pd, Pt) or non noble metal sulphides (Mo, Wo or Co, Ni)] provide hydrogenation dehydrogenation activity. These metals catalyze the hydrogenation of feedstocks making them more reactive for cracking and hetero-atom removal as well reducing the coke rate.

Zeolite based hydrocracking catalysts have following advantages of greater acidity resulting in greater cracking activity; better thermal/hydrothermal stability; better naphtha selectivity; better resistance to nitrogen and sulphur compounds; low coke forming tendency, and easy regenerability.

Once Though Hydrocracking Process

The unit consists of the following sections:

- Furnace
- First stage Reactor section
- Second stage Reactor section
- High pressure separator

- Fractionation Section

- Light Ends Recovery section

In single stage process both treating and cracking steps are combined in a single reactor. Single stage hydrocracking flow diagram is show in Figure [Mall,2007].

In this process the feed along with recycle unconverted residue from the fractionator is first hydro-treated in a reactor and then the combined stream gases are fed to second reactor where cracking takes place in the presence of hydro-cracking catalyst.

In the single stage process the catalysts work under high H_2S and NH_3 partial pressure.

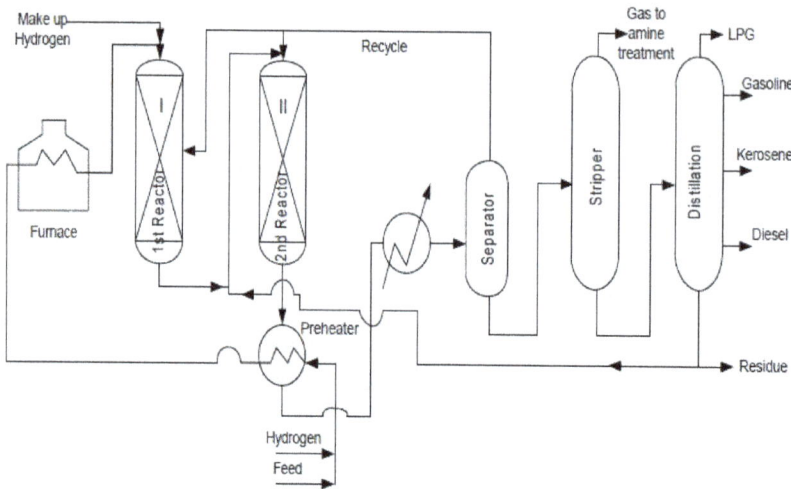

Once Through Hydrocracking

Two Stage Hydrocracking Process

The unit consists of the following sections:

- Furnace

- First stage Reactor section.

- Second stage Reactor section

- Third stage reactor

- Fractionation Section

- Light Ends Recovery section

Preheated feed is first hydro treated in a reactor for desulphurization and denitrogenation in presence of pretreat catalyst followed by hydrocracking in second reactor in presence of strongly acidic catalyst with a relatively low hydrogenation activity. In the first stage reactor the sulphur and nitrogen compounds are converted to hydrogen sulphide and ammonia with limited hydrocarcking. The two stages process employs interstage product separation that removes H_2S and NH_3. In case of two stage process, hydrocracking catalyst works under low H_2S and NH_3. Process flow diagram for two stage hydrocracking process is shown in Figure [mall 2007].

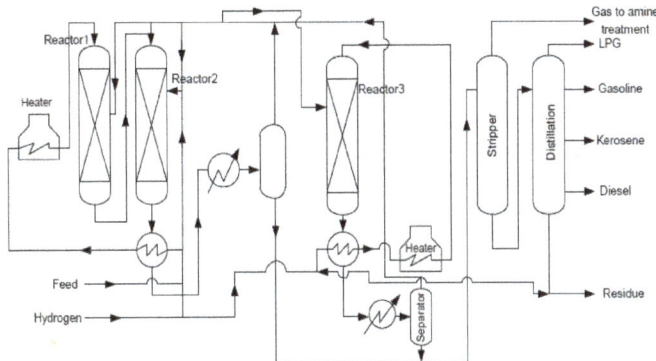

Two Stage Hydrocracking Process

Hydrocracking Chemistry

Hydrocracking process is a catalytic cracking process which takes place in the presence of an elevated partial pressure of hydrogen and is facilitated by bio-functional catalyst having acidic sites and metallic sites. During hydrocracking process hydrotreating reactions and hydrocracking reactions are two major reactions which take place.

A typical hydrocracking reaction is as follows.

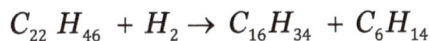

$$C_{22}H_{46} + H_2 \rightarrow C_{16}H_{34} + C_6H_{14}$$

Various hydrotreating reactions are hydrodesulphurization, denitrogenation, hydro deoxygenation, hydro metallization, olefin hydrogenation partial aromatics saturation.

Various hydrocracking reactions are splitting of C-C bond and or C-C rearrangement reaction (hydrisomerisation process) Hydrogenation and dehydrogenation catalyst.

Effect of Operating Parameters

Various operating parameters affecting hydrocracking are

- Reaction temperature
- Hydrogen partial pressure
- Hourly feed velocity of the feed
- Hydrogen recycle ratio

Temperature increase in temperature accelerates cracking reaction on acid sites and displaces the equilibrium of hydrogenation reactions towards dehydrogenation. Too high temperature limits the hydrocracking of aromatic structure .

The pressure influences significantly the equilibrium of dehydrogenation-hydrogenation reactions that takes place on the metallic sites. The increase in pressure for a given molar ratio H_2/feed corresponds to increase in the partial pressure of hydrogen, will produce an increase the conversion of the aromatic structures to saturated products which will improve the quality of jet fuel, diesel fuel and oil with very high viscosity index.

Effect of Feedstock: a higher content of aromatic hydrocarbons requires higher pressure and higher hydrogen/feed ratio, the lowest possible temperature and a higher hydrogen consumption of hydrogen and the severity of the process

Effects of Feed Impurities: Hydrogen sulphide, nitrogen compounds and aromatic molecules present in the feed affect the hydrocracking reactions. Increase in nitrogen result in lower conversion.

Ammonia inhibits the hydrocracking catalyst activity, requiring higher operating temperatures. Polymeric compounds have substantial inhibiting and poisoning effects. Polynuclear aromatics

Effect of various parameters on catalyst life:

Variable	Change Effect on catalyst life
Feed rate	Increase-Decrease
Conversion	Increase-Decrease
Hydrogen partial pressure	Increase-Increase
Reactor pressure	Increase-Increase
Recycle gas rate	Increase -Increase
Recycle gas purity	Increase -Increase

Hydro-Cracking Technology Provider

- Chevron: Isocracking
- UOP: Uni-cracking
- IFP: Hydrocarcking
- B.P. U.K: Hydrocracking
- Shell: Hydrocracking
- Standard Oil: Ultracracking

- Linde: Hydrocarking

- Union Oil Co.: Uni-cracking

Catalyst Deactivation

Catalyst activation may occur due to coke deposition and metal accumulation.

Coke Depositions may be due to condensation of poly-nuclear and olefinic compounds into high molecular weight which cover active sites. Metal Accumulation occurs at the pore entrances or near the outer surface of the catalyst

Catalyst Regeneration

Catalyst regeneration is done by burning off the carbon, and sulphur and circulation of circulate nitrogen with the recycle compressor, injecting a small quantity of air and maintaining catalyst temperature above the coke ignition temperature.

Catalytic Reforming

Catalytic reforming is a chemical process used to convert petroleum refinery naphthas distilled from crude oil (typically having low octane ratings) into high-octane liquid products called reformates, which are premium blending stocks for high-octane gasoline. The process converts low-octane linear hydrocarbons (paraffins) into branched alkanes (isoparaffins) and cyclic naphthenes, which are then partially dehydrogenated to produce high-octane aromatic hydrocarbons. The dehydrogenation also produces significant amounts of byproduct hydrogen gas, which is fed into other refinery processes such as hydrocracking. A side reaction is hydrogenolysis, which produces light hydrocarbons of lower value, such as methane, ethane, propane and butanes. It is also the conversion of straight chains of alkane catalytically

In addition to a gasoline blending stock, reformate is the main source of aromatic bulk chemicals such as benzene, toluene, xylene and ethylbenzene which have diverse uses, most importantly as raw materials for conversion into plastics. However, the benzene content of reformate makes it carcinogenic, which has led to governmental regulations effectively requiring further processing to reduce its benzene content.

This process is quite different from and not to be confused with the catalytic steam reforming process used industrially to produce products such as hydrogen, ammonia, and methanol from natural gas, naphtha or other petroleum-derived feedstocks. Nor is this process to be confused with various other catalytic reforming processes that use methanol or biomass-derived feedstocks to produce hydrogen for fuel cells or other uses.

History

In the 1940s, Vladimir Haensel, a research chemist working for Universal Oil Products (UOP), developed a catalytic reforming process using a catalyst containing platinum. Haensel's process was subsequently commercialized by UOP in 1949 for producing a high octane gasoline from low octane naphthas and the UOP process become known as the Platforming process. The first Platforming unit was built in 1949 at the refinery of the Old Dutch Refining Company in Muskegon, Michigan.

In the years since then, many other versions of the process have been developed by some of the major oil companies and other organizations. Today, the large majority of gasoline produced worldwide is derived from the catalytic reforming process.

To name a few of the other catalytic reforming versions that were developed, all of which utilized a platinum and/or a rhenium catalyst:

- Rheniforming: Developed by Chevron Oil Company.

- Powerforming: Developed by Esso Oil Company, currently known as ExxonMobil.

- Magnaforming: Developed by Engelhard and Atlantic Richfield Oil Company.

- Ultraforming: Developed by Standard Oil of Indiana, now a part of the British Petroleum Company.

- Houdriforming: Developed by the Houdry Process Corporation.

- CCR Platforming: A Platforming version, designed for continuous catalyst regeneration, developed by UOP.

- Octanizing: A catalytic reforming version developed by Axens, a subsidiary of Institut francais du petrole (IFP), designed for continuous catalyst regeneration.

Chemistry

Before describing the reaction chemistry of the catalytic reforming process as used in petroleum refineries, the typical naphthas used as catalytic reforming feedstocks will be discussed.

Typical Naphtha Feedstocks

A petroleum refinery includes many unit operations and unit processes. The first unit operation in a refinery is the continuous distillation of the petroleum crude oil being refined. The overhead liquid distillate is called naphtha and will become a major component of the refinery's gasoline (petrol) product after it is further processed through a catalytic hydrodesulfurizer to remove sulfur-containing hydrocarbons and a catalytic reformer to reform its hydrocarbon molecules into more complex molecules with a higher octane rating value. The naphtha is a mixture of very many different hydrocarbon compounds. It has an initial boiling point of about 35 °C and a final boiling point of about 200 °C, and it contains paraffin, naphthene (cyclic paraffins) and aromatic hydrocarbons ranging from those containing 4 carbon atoms to those containing about 10 or 11 carbon atoms.

The naphtha from the crude oil distillation is often further distilled to produce a "light" naphtha containing most (but not all) of the hydrocarbons with 6 or fewer carbon atoms and a "heavy" naphtha containing most (but not all) of the hydrocarbons with more than 6 carbon atoms. The heavy naphtha has an initial boiling point of about 140 to 150 °C and a final boiling point of about 190 to 205 °C. The naphthas derived from the distillation of crude oils are referred to as "straight-run" naphthas.

It is the straight-run heavy naphtha that is usually processed in a catalytic reformer because the light naphtha has molecules with 6 or fewer carbon atoms which, when reformed, tend to crack into butane and lower molecular weight hydrocarbons which are not useful as high-octane gasoline blending components. Also, the molecules with 6 carbon atoms tend to form aromatics which is undesirable because governmental environmental regulations in a number of countries limit the amount of aromatics (most particularly benzene) that gasoline may contain.

It should be noted that there are a great many petroleum crude oil sources worldwide and each crude oil has its own unique composition or "assay". Also, not all refineries process the same crude oils and each refinery produces its own straight-run naphthas with their own unique initial and final boiling points. In other words, naphtha is a generic term rather than a specific term.

The table just below lists some fairly typical straight-run heavy naphtha feedstocks, available for catalytic reforming, derived from various crude oils. It can be seen that they differ significantly in their content of paraffins, naphthenes and aromatics:

Typical Heavy Naphtha Feedstocks				
Crude oil name Location	**Barrow Island Australia**	**Mutineer-Exeter Australia**	**CPC Blend Kazakhstan**	**Draugen North Sea**
Initial boiling point, °C	149	140	149	150
Final boiling point, °C	204	190	204	180
Paraffins, liquid volume %	46	62	57	38
Naphthenes, liquid volume %	42	32	27	45
Aromatics, liquid volume %	12	6	16	17

Some refinery naphthas include olefinic hydrocarbons, such as naphthas derived from the fluid catalytic cracking and coking processes used in many refineries. Some refineries may also desulfurize and catalytically reform those naphthas. However, for the most part, catalytic reforming is mainly used on the straight-run heavy naphthas, such as those in the above table, derived from the distillation of crude oils.

The Reaction Chemistry

There are many chemical reactions that occur in the catalytic reforming process, all of which occur in the presence of a catalyst and a high partial pressure of hydrogen. Depending upon the type or version of catalytic reforming used as well as the desired reaction severity, the reaction conditions range from temperatures of about 495 to 525 °C and from pressures of about 5 to 45 atm.

The commonly used catalytic reforming catalysts contain noble metals such as platinum and/or rhenium, which are very susceptible to poisoning by sulfur and nitrogen compounds. Therefore, the naphtha feedstock to a catalytic reformer is always pre-processed in a hydrodesulfurization unit which removes both the sulfur and the nitrogen compounds. Most catalysts require both sulphur and nitrogen content to be lower than 1 ppm.

The four major catalytic reforming reactions are:

1: The dehydrogenation of naphthenes to convert them into aromatics as exemplified in the conversion methylcyclohexane (a naphthene) to toluene (an aromatic), as shown below:

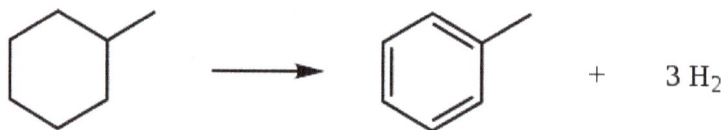

$$+ \quad 3 H_2$$

2: The isomerization of normal paraffins to isoparaffins as exemplified in the conversion of normal octane to 2,5-Dimethylhexane (an isoparaffin), as shown below:

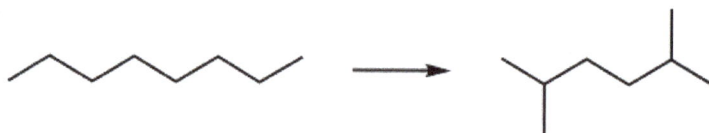

3: The dehydrogenation and aromatization of paraffins to aromatics (commonly called dehydrocyclization) as exemplified in the conversion of normal heptane to toluene, as shown below:

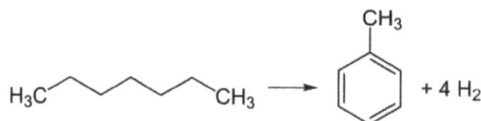

$$+ 4 H_2$$

4: The hydrocracking of paraffins into smaller molecules as exemplified by the cracking of normal heptane into isopentane and ethane, as shown below:

n-Heptane + H_2 $\longrightarrow$ Isopentane + Ethane

$$H-\underset{\underset{H}{|}}{\overset{\overset{H}{|}}{C}}-\underset{\underset{H}{|}}{\overset{\overset{H}{|}}{C}}-\underset{\underset{H}{|}}{\overset{\overset{H}{|}}{C}}-\underset{\underset{H}{|}}{\overset{\overset{H}{|}}{C}}-\underset{\underset{H}{|}}{\overset{\overset{H}{|}}{C}}-\underset{\underset{H}{|}}{\overset{\overset{H}{|}}{C}}-\underset{\underset{H}{|}}{\overset{\overset{H}{|}}{C}}-H \; + \; H_2 \; \longrightarrow \; H-\underset{\underset{H}{|}}{\overset{\overset{H}{|}}{C}}-\underset{\underset{H}{|}}{\overset{\overset{CH_3}{|}}{C}}-\underset{\underset{H}{|}}{\overset{\overset{H}{|}}{C}}-\underset{\underset{H}{|}}{\overset{\overset{H}{|}}{C}}-H \; + \; H-\underset{\underset{H}{|}}{\overset{\overset{H}{|}}{C}}-\underset{\underset{H}{|}}{\overset{\overset{H}{|}}{C}}-H$$

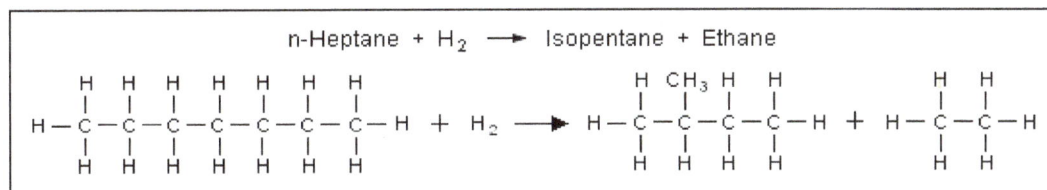

The hydrocracking of paraffins is the only one of the above four major reforming reactions that consumes hydrogen. The isomerization of normal paraffins does not consume or produce hydrogen. However, both the dehydrogenation of naphthenes and the dehydrocyclization of paraffins produce hydrogen. The overall net production of hydrogen in the catalytic reforming of petroleum naphthas ranges from about 50 to 200 cubic meters of hydrogen gas (at 0 °C and 1 atm) per cubic meter of liquid naphtha feedstock. In the United States customary units, that is equivalent to 300 to 1200 cubic feet of hydrogen gas (at 60 °F and 1 atm) per barrel of liquid naphtha feedstock. In many petroleum refineries, the net hydrogen produced in catalytic reforming supplies a significant part of the hydrogen used elsewhere in the refinery (for example, in hydrodesulfurization processes). The hydrogen is also necessary in order to hydrogenolyze any polymers that form on the catalyst.

In practice, the higher the content of naphtenes in the naphtha feedstock, the better will be the quality of the reformate and the higher the production of hydrogen. Crude oils containing the best naphtha for reforming are typically from Western Africa or the North Sea, such as Bonny light or Troll.

Process Description

The most commonly used type of catalytic reforming unit has three reactors, each with a fixed bed of catalyst, and all of the catalyst is regenerated *in situ* during routine catalyst regeneration shutdowns which occur approximately once each 6 to 24 months. Such a unit is referred to as a semi-regenerative catalytic reformer (SRR).

Some catalytic reforming units have an extra *spare* or *swing* reactor and each reactor can be individually isolated so that any one reactor can be undergoing in situ regeneration while the other reactors are in operation. When that reactor is regenerated, it replaces another reactor which, in turn, is isolated so that it can then be regenerated. Such units, referred to as *cyclic* catalytic reformers, are not very common. Cyclic catalytic reformers serve to extend the period between required.

The latest and most modern type of catalytic reformers are called continuous catalyst regeneration (CCR) reformers. Such units are characterized by continuous in-situ regeneration of part of the catalyst in a special regenerator, and by continuous addition of the regenerated catalyst to the operating reactors. As of 2006, two CCR versions available: UOP's CCR Platformer process and Axens' Octanizing process. The installation and use of CCR units is rapidly increasing.

Many of the earliest catalytic reforming units (in the 1950s and 1960s) were non-regenerative in that they did not perform in situ catalyst regeneration. Instead, when needed, the aged catalyst was replaced by fresh catalyst and the aged catalyst was shipped to catalyst manufacturers to be either regenerated or to recover the platinum content of the aged catalyst. Very few, if any, catalytic reformers currently in operation are non-regenerative.

The process flow diagram below depicts a typical semi-regenerative catalytic reforming unit.

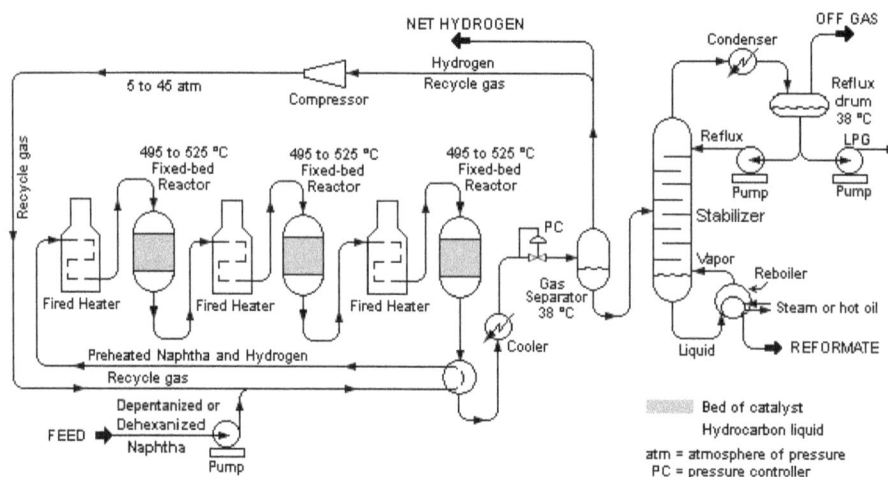

Schematic diagram of a typical semi-regenerative catalytic reformer unit in a petroleum refinery

The liquid feed (at the bottom left in the diagram) is pumped up to the reaction pressure (5–45 atm) and is joined by a stream of hydrogen-rich recycle gas. The resulting liquid–gas mixture is preheated by flowing through a heat exchanger. The preheated feed mixture is then totally vaporized and heated to the reaction temperature (495–520 °C) before the vaporized reactants enter the first reactor. As the vaporized reactants flow through the fixed bed of catalyst in the reactor, the major reaction is the dehydrogenation of naphthenes to aromatics (as described earlier herein) which is highly endothermic and results in a large temperature decrease between the inlet and outlet of the reactor. To maintain the required reaction temperature and the rate of reaction, the vaporized stream is reheated in the second fired heater before it flows through the second reactor. The temperature again decreases across the second reactor and the vaporized stream must again be reheated in the third fired heater before it flows through the third reactor. As the vaporized stream proceeds through the three reactors, the reaction rates decrease and the reactors therefore become larger. At the same time, the amount of reheat required between the reactors becomes smaller. Usually, three reactors are all that is required to provide the desired performance of the catalytic reforming unit.

Some installations use three separate fired heaters as shown in the schematic diagram and some installations use a single fired heater with three separate heating coils.

The hot reaction products from the third reactor are partially cooled by flowing through the heat exchanger where the feed to the first reactor is preheated and then flow through a water-cooled heat exchanger before flowing through the pressure controller (PC) into the gas separator.

Most of the hydrogen-rich gas from the gas separator vessel returns to the suction of the recycle hydrogen gas compressor and the net production of hydrogen-rich gas from the reforming reactions is exported for use in the other refinery processes that consume hydrogen (such as hydrodesulfurization units and/or a hydrocracker unit).

The liquid from the gas separator vessel is routed into a fractionating column commonly called a *stabilizer*. The overhead offgas product from the stabilizer contains the byproduct methane, ethane, propane and butane gases produced by the hydrocracking reactions as explained in the above discussion of the reaction chemistry of a catalytic reformer, and it may also contain some small amount of hydrogen. That offgas is routed to the refinery's central gas processing plant for removal and recovery of propane and butane. The residual gas after such processing becomes part of the refinery's fuel gas system.

The bottoms product from the stabilizer is the high-octane liquid reformate that will become a component of the refinery's product gasoline. Reformate can be blended directly in the gasoline pool but often it is separated in two or more streams. A common refining scheme consists in fractionating the reformate in two streams, light and heavy reformate. The light reformate has lower octane and can be used as isomerization feedstock if this unit is available. The heavy reformate is high in octane and low in benzene, hence it is an excellent blending component for the gasoline pool.

Benzene is often removed with a specific operation to reduce the content of benzene in the reformate as the finished gasoline has often an upper limit of benzene content (in the UE this is 1% volume). The benzene extracted can be marketed as feedstock for the chemical industry.

Catalysts and Mechanisms

Most catalytic reforming catalysts contain platinum or rhenium on a silica or silica-alumina support base, and some contain both platinum and rhenium. Fresh catalyst is chlorided (chlorinated) prior to use.

The noble metals (platinum and rhenium) are considered to be catalytic sites for the dehydrogenation reactions and the chlorinated alumina provides the acid sites needed for isomerization, cyclization and hydrocracking reactions. The biggest care has to be exercised during the chlorination. Indeed, if not chlorinated (or insufficiently chlorinated) the platinum and rhenium in the catalyst would be reduced almost immediately to metallic state by the hydrogen in the vapour phase. On the other an excessive chlorination could depress excessively the activity of the catalyst.

The activity (i.e., effectiveness) of the catalyst in a semi-regenerative catalytic reformer is reduced over time during operation by carbonaceous coke deposition and chloride loss. The activity of the catalyst can be periodically regenerated or restored by in situ high temperature oxidation of the coke followed by chlorination. As stated earlier herein, semi-regenerative catalytic reformers are regenerated about once per 6 to 24 months. The higher the severity of the reacting conditions (temperature), the higher is the octane of the produced reformate but also the shorter will be the duration of the cycle between two regenerations. Catalyst's cycle duration is also very dependent on the quality of the feedstock. However, independently of the crude oil used in the refinery, all catalysts require a maximum final boiling point of the naphtha feedstock of 180 °C.

Normally, the catalyst can be regenerated perhaps 3 or 4 times before it must be returned to the manufacturer for reclamation of the valuable platinum and/or rhenium content.

Alkylation

Alkylation process commercialized in 1938, since then there has been tremendous growth in the process. In US and Europe about alkylate is about 11-12 percent and 6 percent in the gasoline pool respectively. Alkylate is a key component in reformulated gasoline. Alkylation processes are becoming important due to growing demand for high octane gasoline and requirement of low RVP, low sulphur, low toxics. Alkylate is an ideal blend stock to meet these requirement.

The process of alkylation different iso-parraffins using olefins were developed during thirties using aluminium chloride catalyst, however, later catalyst was replaced by HF and sulfuric acid. Although butylenes alkylation is one of the most commonly used process, however, alkylation of amylenes obtained from C_5 fraction of FCC can be another route to increase the availability of alkylate. Alkylation of C_5 cut from FCC can significantly reduce RVP of finished gasoline pool.

C_5 alkylate: Amylene alkylation has two fold advantage: It increase the volume of alkylate available while decreasing Reid vapor pressure and olefinic content of gasoline blend stocks.

The process of HF alkylation produces high octane blend stock for iso-parraffin (mainly iso butane) and olefin (propylene, butylene and amylenes) in the process of HF catalyst to meet all the criteria of reformulated gasoline. Replacing high risk toxic liquid acids, such as hydrofluoric acid (HF) and sulphuric acid with solid acid catalysts is challenging goal iso-parraffin alkylation technology.

Process

The reaction involved in aliphatic alkylation consists of conversion of iso-butane and butylenes to iso-octanes using HF catalyst. Commonly alkylation process used are mention in Table.

$$i-C_4H_4+i-C_4H_8 \rightarrow C_8H_{18}$$

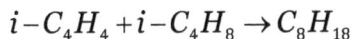

The side reaction results in increased iso-butane consumption increased acid consumption increased acid soluble formation, equipment handling and for the corrosion problem. Figure gives the details of iso-parraffin alkylation mechanism.

Some of the other side reaction is the formation of paraffin, which boils above and below the desired product. Impurities in the feed acid and normal operating practices all can contribute to additional side reactions. Comparison of Alkyclean Technology with Modern Sulphuric Acid and Hydrofluoric Acid Technologies is shown in Table.

The key factors to be controlled in alkylation process are

- Maintaining proper composition of reaction mixture which include isobutene olefins and the HF acid

- Maintaining the proper reaction environment which includes adequate contacting, controlled temperature, and freedom from surges.

- Making a proper separation of the reactor effluent into its various components

Table : Common Alkylation Processes

Process	Description
CONOCO Phillips process (ReVA Process)	Alkylation of propylene, butylenes, pentenes and isobutane to high quality motor fuel using HF catalyst
Stratco INC	Alkylation of propylene, butylenes and amylenes with isobutane using strong sulfuric acid to produce high octane branched chain hydrocarbons using effluent refrigeration alkylation process
UOP HF Alkylation Process	Alkylation of isobutane with light olefins (propylene, butylenes and amylenes to produce branched chain parafinic fuel) using hydrofluoric acid catalyst. More than 100 commercial process
UOP Alkylene	UOP Alkylene process is based on solid catalyst(HAL-100) for alkylation of light olefins and isobutane to form a complex mixture of isoalkanes which are highly branched trimethylpentanes(TMP) that have high octane blend values of approximately100
Exxon Alkylation	Alkylation of propylene, butylenes and penrylene with isobutene in the presence of sulphuric acid catalyst using autorefrigeration. Products: a low sensitivity, highly iso, low RVP, high octane gasoline blend stock paraffinic
AlkylClean solid Acid alkylation technology (ABBLumus global)	The alkylation process uses a robust zeolite solid acid catalyst formulation coupled with a novel reactor processing scheme to yield a high quality alkylate product. The catalyst contains no halogen

Table : Comparison of Alkyclean Technology with Modern Sulphuric Acid and Hydro-fluoric Acid Technologies

Parameter	Modern sulphuric acid technology	Modern hydrofluoric acid technology	Alkyclean
Base condition	C4 feedstock	C4=feedstock	C4=feedstock
Product RON	95	95	95
Product MON	Base	Base or better	Base or better
Alkylate yield	Base	BASE	90% of base
Total installed cost	Base	85% OF BASE	50% of base
Total installed cost, including OSBL(regeneration, facilities, and /or safety installations)	Base	Less	None
ASO yield	Base	Less	None
Equipment maintenance	High	High	Much lower
Corrosion problems	Yes	Yes	Higher
reliability and on stream factor	Base	Base	Match fcc or better/ shorter
Safety	Unit specific safety precautions as well as transport precautions unit specific precautions	C safety precautions required that extend throughout refinery very specific	No special precautions other than those for any refinery process unit
Catalyst	H2SO4	HF	Zeolite
Environmental	Significant waste streams generated	Significant waste streams generated	No emissions to air, water, or ground

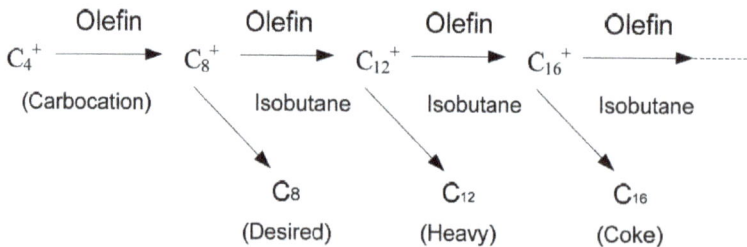

Iso-paraffin Alkylation Mechanism

Isomerisation

Petroleum fractions contain significant amounts of n-alkanes and the isomerisation of al-kanes into corresponding branched isomers is one of the important process in refining . The highly branched paraffins with 7-10 carbon atoms would be the best to fulfill the recent requirements of the reformulated gasoline. The production of paraffin bases high – octane gasoline blend stock, such as isomers from isomerisation of light and mid cut naphtha might be a key technology for gasoline supply to cope with future gasoline regulation.

Light naphtha and paraffin isomerisation recognizes emerging technologies in order to boost octane in light gasoline fractions. Recent pricing trends show isomerisation could be a significant contributor to octane pool which will offset the loss from gasoline de-sulfurisation and aromatic reduction. Isomerate as % of gasoline used is USA 8percent, Western Europe 16percent. Isomerisation involves.

- Isomerisation of Light paraffins
- Isomerisation of C_5-C_6 paraffins
- Isomerisation of n-butane

Isomerisation of C_5-C_6 paraffins: Allow low octane number paraffins 5 and 6 carbon atoms into higher octane number paraffins.

n-pentanen-isopentane:

n-hexane to 2-methyl pentane, 3 methyl pentane (low octane 75) 2,2 dimethyl butane, 2,3 dimethyl butane.

Isomerisation of n-Butane; to produce isobutene feed for alkylation or as source of isobutene dehydrogenation to manufacture MTBE.

Isomerisation Catalyst

Two types of isomerization catalyst, zeolite and chlorinated alumina, has been used. Zeolite catalyst requires higher temperatures and provide lower octane boost while chlorinated alumina's results in highest octane, however, it has higher sensitivity to feed stock impurities requiring strict feed pretreatment to eliminate oxygen, water, sulphur and nitrogen is containing compounds.

- Zeolite
- Chlorinated alumina

Zeolite catalyst requires higher temperatures and provide lower octane boost

Chlorinated alumina's results highest octane, however, it has higher sensitivity to feed stock impurities requiring strict feed pretreatment to eliminate oxygen, water , sulphur and nitrogen containing compounds.

Isomerisation of Light Naphtha

C_5/C_6 feed either from straight run crude distillation or from catalytic reforming. Table gives details of isomerization of light paraffins catalyst.

Reformate: separated in lighter mostly benzene and heavier containing C7 Catalyst: Zeolite or Pt on Chlorinated alumina

Operating Condition:

	Pt on chlorinated alumina	Pt on zeolite
Temperature oC	120-180	250-270
Pressure	20-30	15-30
Space velocity h^{-1}	1-2	1-2
H$_2$/HC ratio	0.1-2	2-4
Product RON	83-84	78-80

Once Through Process

Recycle Process: Unconverted n-paraffins and any single branched isomers from double branched isomers

Recycling with Distillation: Deisohexaniser

Recycling with Adsorption: Adsorption on Molecular sieve: n-paraffins are adsorbed and separated by desorption

Table : Isomerisation of Light Paraffins Catalyst

Isomerisation Catalyst	
1st generation	Friedel and Crafts AlCl3 catalysts, exhibit very high activity at low temp980-100°C
2nd generation	Metal/ support bifunctional catalyst essentially Pt/alumina sensitivity to poisons are less acute, however, require higher temperature (350-550°C.
3rd generation	Metal/support bifunctional catalysts with increased acidity by halogenation of the alumina support. Sensitive to poisons and need pretreatment, Corrosion problem. High activity at low temperature9120°C-to 160°C
4th generation	Bifunctional zeolite catalysts, very resistant to catalyst poison and feed does not need pretreatment

Isomerisation of n-butane

To produce isobutene feed for alkylation or as source of isobutene dehydrogenation to manufacture MTBE

UOP Butamar Process:

Catalyst: Pt/chlorintated Al$_2$O$_3$

Operating Condition: Temperature: 180-220 °C,

Pressure: 15-20 bar

Soacevel: $2h^{-1}$

H2/HC: 0.5 to 2

UOP isomerisation Technologies:

Some of the UOP Light paraffin isomerisation technology are

Penex: Higher octanes, higher product yields more than 120 licensed units

Par – Isom: UOP introduced par-ISOM in 1996 using zeolite chloride sulfate of zirconium catalyst. It is chracterised by lower equipment cost, multiple catalyst approach. Some advantage of Penex process Maximum octane bbls, high octane, best long-term profitability higher investment cost. It can handle undesired feedstocks including feed and process high benzene content feeds. It has wide range of operation. Penex once through Penex plus for extra high benzene levels DIH, DIP/DIH, MDEX

Penex :Para-ISOM process with PI-242 catalyst: Best LPG production , good octane, rapid payback, low investment cost

Polymerisation

Polymerization processes have received considerable interest in petroleum refining because of the higher requirement of reformulated gasoline and phasing of MTBE. The process may be attractive in two main areas [Leprince, 1998].

- Upgrading of C_2 and C Temperature: 150-200°C, Pressure: 30-50bar, space velocity 0.3-0.5 m^3/h per m^3 cuts from catalytic cracking for oligmerization ethylene & propylene to olefinic gasoline.

- Producing high quality middle quality

The level of sulphur in the past two decades has steadily increased due to use of more and more heavier crude, use of cheaper high sulphur crude which has forced the refining industry to go for additional facilities like ultra-desulphuristion for gasoline and diesel to meet the requirement of the stringent sulphur emission standards. Requirement of sulphur content for MS and HSD is given in Table. Sulphur is one of the major impurities in heavy crude resulting higher concentration of sulphur compounds in the un-desulphurised product stream. Sulphur content in the crude varies widely depending on the origin. Table shows sulphur content in crude oil. The variation is considerable and this impacts the processing scheme as well as the product slate [Goel et al.,2008]. Sulphur content of commonly used sweet and sour crudes. Due to increasing environmental concerns, stringent limits on sulphur levels in fuel are being implemented world over to achieve target of sulphur below 100 ppm, deep hydrodesulphurization is required which is an additional capital cost as well as an energy intensive step. Table given the details of reactivity of sulphur compounds present in crude oil.

Table : Sulphur Requirement in Different Gasoline & Diesel in PPM

	BIS 2000	Bharat stage-II	Euro-III equivalent	Euro-IV equivalent
MS	1000	500	150	50
HSD	2500	500	350	50

Table : Sulphur Content in Crude Oil

Crude Name	'S' Content, (wt%)
Bombay High	0.17
Bonny Light	0.14
Arab Heavy	2.87
Arab Light	1.09
Doba	016
Ratawi	3.88
Miri light	0.078
Tapis Blend	0.028

Table : Reactivity of Sulphur Compounds Present in Crude Oil

Sulphur compound	Relative reaction rate	Boiling pounts
Thiophene	100	185
Benzothiphene	50	430
Dibenzothiophene	30	590
Dimethyldibenzothiophene	5	600-620
trimethyldibenzothiophene		630-680

Sulphur Output

Sulphur output from the refinery takes places as one of the following [Goel, 2008]

- Sulphur content in finished product

- Sulphur emission into atmosphere in the form of SO_2

- Sulphur recovery in sulphur recovery unit Sulphur distribution in typical refinery is given in below

- Sulphur in various products 58%

- Product sulphur 41%

- Sulphur emission 1%

Future Demand in Indian Refineries

Use of more sour crudes

More stringent sulphur specifications for distillate products More stringent specifications for SOx emission through flue gas

Leading to enhancement in sulphur recovery and capacity augmentation

Estimated sulphur recovery capacity in Indian Refineries to be more than double in near future

Environmental regulation shave been the major driving force for reducing sulphur in refinery products.

Process Used to Remove Sulphur from Different Products

- LPG - LPG treating unit
- Gasoline - Hydrotreating Unit
- ATF - Merox / Hydrotreating
- Diesel - Hydrotreating
- Sulfur lands up in the fuel gas as H_2S during Hydrotreating
- H_2S in fuel gas produces SOx while burning in the fired heater
- Environmental Norm for SOx : 50 ppm while burning fuel gas

Hydro Treatment Processes

Hydro treatment of the various streams from refinery and petrochemical industries has become integral part in order to meet the feed standards of various processes in order to avoid catalyst poisoning, improving quality of products and meet the environmental standards.

Hydroprocessing technologies consist of any one of the following processes

Pretreatment (Hydrotreatment) of naphtha and gas oil, residue for Catalytic reforming, Catalytic cracking and Hydrocracking in order to remove the impurities sulphur, nitrogen, heavy metal etc.

Hydrocracking Processes

Hydrotreatment of the fuels and lubricants, Hydro treatment of naphtha, gas oil and residue for catalytic reforming, catalytic cracking and hydrocracking processes. These processes have been discussed separately in the chapters catalytic reforming catalytic cracking and hydrocracking. Various hydrotreament processes removes sulphur compound which must be recovered in sulphur recovery units. Figure illustrates the major sources of sulphur & recovery processes in refinery.

Main reaction involved in desulphurization is removal of sulphur compounds in form of H_2S. Degree of desulphurization varies from feed to feed with nearly complete removal to about 50- 70percent for heavier residual materials.

Relative desulphurization reactivity is in order of increasing difficulty is given below:

Thiphenol> ethyl mercaptan> diethyl sulphide>diphenylsulphide>3-metrhyl-1butanethiol>diethylsulphide>di[ropylsulphide>diisomylsulphide>thiophene

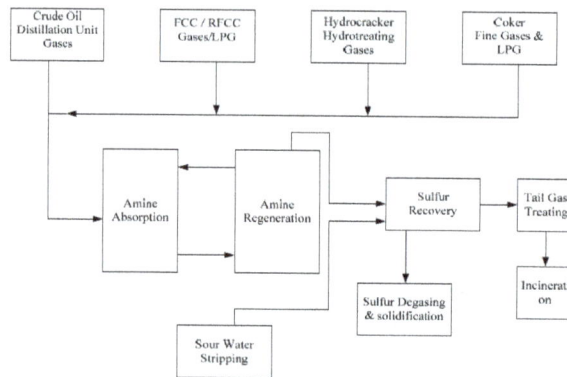

Major sources of Sulphur & Recovery Processes in Refinery

Sulphur Recovery Units Characteristics – Refineries

- Small to Medium Size Sulphur Recovery Units

- From a few tons to a few hundred tons/day

- Guwahati Refinery, IOCL : 5 TPD

- Reliance Refinery, Jamnagar : 2025 (3 x 675) TPD

- Feed composition varies, linked to Refinery operating mode and Crude feedstock

- High flexibility required, multiple trains

- Acid Gas always rich (high H_2S content)

- Ammonia (from Sour Water Stripper) always present, sometimes in relatively high quantities

Sulphur Recovery Unit

Sulphur recovery unit consist of recovery of sulphur from H_2S present in acid gas from Amine Treating/ Regeneration unit and H_2S from sour water stripper section Hydrogen sulphide content of the feed gas is converted to elemental sulphur. Typical sulphur recovery unit is shown in Figure.

Amine absorption and Regeneration: Absorption of H_2S bearing stream and regeneration of amine. H_2S rich stream from amine regeneration is sent to sulphur recovery unit.

Sour Water Stripping: Sour water is tripped off its sulphur and recycled. H_2S is sent to sulphur recovery unit.

Amine Absorption Unit: Various hydro desuplhurisation processes in the refinery and hydrocracker unit generate large quantity of H_2S. H_2S bearing gases from various unit is sent to Amine treating unit which uses amine as a solvent for absorbing H_2S and subsequently releasing H_2S as H_2S Rich stream in the amine generator.

Merox (Mercaptan Oxidation Unit)

Merox process is used in the refinery for controlling the mercaptan sulphur in gases, LPG, naphtha and other petroleum fractions. The Process is used for the chemical treatment of LPG, gasoline and distillates from FCCU, OHCU etc to remove mercaptans.

Mercaptans are either extracted from the stream or sweetened to acceptable disulphides. For treatment of light feed stocks such as LPG, no sweetening is required as mercaptans are nearly removed by extraction. However, feed containing higher molecular weight mercaptans and may require a combination of Merox extraction and sweetening using catalyst. Catalysts promote the oxidation of mercaptans to disulphide using air as the source of oxygen. Merox treatment can in general be used in following ways.

- To improve lead susceptibility of light gasoline
- To improve the response of gasoline stocks to oxidation inhibitors added to prevent gum formation during storage
- To improve odor on all stocks
- To reduce the mercaptans content to meet product specifications
- To reduce the sulphur content of LPG and light naphtha products
- To reduce sulphur content of coker FCC olefins to save acid consumption in alkylation

Process

$$RSH + NaOH \rightarrow RSNa + H_2O$$

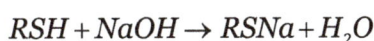

Pretreatment(Remove H_2S and Naphthenic Acids by dilute Alkali Solution)

Extraction (Remove Caustic soluble Mercaptans)

Sweetening(Oxidation of mercaptans to disulphides)

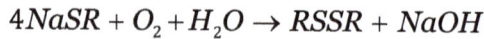

$$4NaSR + O_2 + H_2O \rightarrow RSSR + NaOH$$

Post Treatment (Remove Caustic Haze)

(Caustic Settler, Wash Water, Sand, Clay Filters

Sulphur Recovery from H$_2$S

Sulphur recovery now has become one of the most critical aspects of sulphur management and affects emission sulphur dioxides significantly in the refinery. There are two sulphur recovery processes are

- Claus process(used earlier)
- Supr Claus process

Conventional Claus process has only 99% sulphur recovery. In order to meet the sulphur emission standards now Claus process has been improved substantially to meet the standards. Modern claus process is shown in Figure. New processes are characterized by

- New Catalysts
- COS and CS$_2$ hydrolysis (increased recovery)
- Direct conversion of H$_2$S to Sulphur by oxidation (Super Claus Process)
- Direct conversion of H$_2$S to Sulphur by reduction (Pro Claus Process)
- High efficiency burners (NH3, BTEX destruction)
- Analysers based control
- Enriched air or Oxygen blown thermal reactors

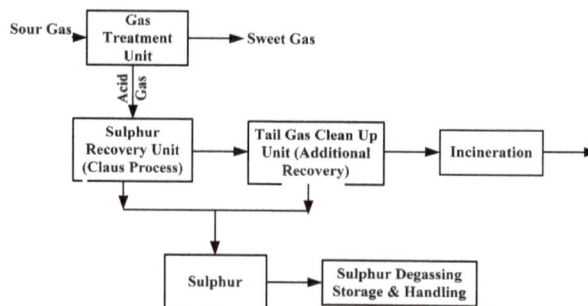

Typical Sulphur Recovery

Modified Claus Process

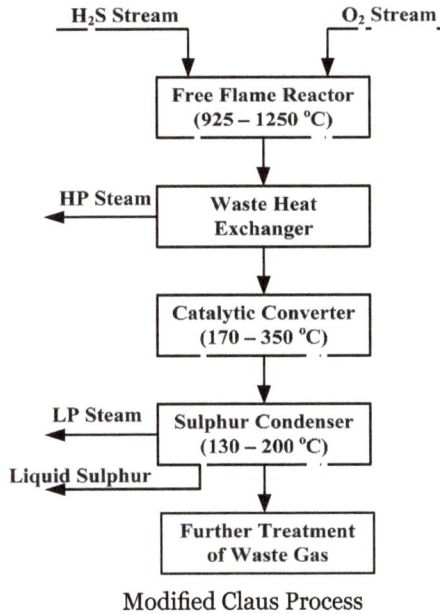

Super Claus Process

The SUPER CLAUS process was developed to catalytically recover elemental sulphur from H_2S containing Claus tail gas to improve the overall sulphur recovery level. The SUPERCLAUS process was commercially demonstrated in 1998, and today now more than 160 units are under license and over 140 are in operation. SUPERCLAUS process achieves high sulphur recover levels by suppressing SO_2 formation in claus stages and selectively oxidizing H_2S in presence of oxygen using proprietary catalyst.

A typical SUPER CLAUS sulphur recovery unit consist of following sections:

- Combustion Chamber

- Claus reactor

- Super claus Reactor

- Incinerator

- Degassing Section

Function of Claus reactors:

- Claus reaction at catalytic region

$$2H_2S + SO_2 \rightarrow 3/xS_x + 2H_2O + 93k \, (Where \; x = 6 \; and \; 8 \; mainly)$$

- Hydrolysis of COS and CS_2 at temperatures above $300°C$

$$COS + H_2O \rightarrow CO_2 + H_2S$$

$$CS_2 + 2H_2O \rightarrow CO_2 + 2H_2S$$

Function of Super Claus Reactor

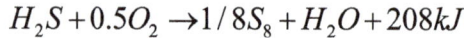

$$H_2S + 0.5O_2 \rightarrow 1/8S_8 + H_2O + 208kJ$$

SUPER CLAUS Process use selective oxidation catalyst minimizes side reactions & increase sulphur recovery

Claus Process Limitations:

- Thermodynamically limited conversion: $2H_2S + SO_2 \rightarrow 3S + H_2O$ the 'air to clean gas' ratio's is maintained to produce an H_2S/SO_2 ratio of exactly 2/1 (optimum ratio) in the burner effluent gases.

- Increases H_2O content to 30 vol% decreasing H_2S and SO_2 concentrations.

Formation of non-recoverable S-compounds due to side reactions

The big difference between SUPER CLAUS catalyst and Claus Catalyst is that the reaction is not equilibrium based. Therefore, the conversion efficiency is much higher than the equilibrium limited Claus reaction. SUPER CLAUS is a non-cyclic process that has repeatedly shown simplicity in operation, high online reliability and sulphur guarantees up to 99.3percent.

Super Sour Process: Stringent environmental regulations have necessitated higher recovery of H_2S from sour water stripper unit design. Super Sour process ensures minimum H_2S loss. the process employ additional hot feed flash drum upstream of cold feed surge drum. The H_2S rich vapours from hot feed flash drum upstream of cold feed surge drum is routed to a small amine scrubber to absorb liberated H_2S. The H_2S lean gas containing primarily hydrocarbons is then routed to incinerator of the sulphur recovery unit. The absorbed H_2S rich amine is recovered in the amine regenerator and is fed to the sulphur unit for converting it to sulphur [Sharma and Nag 2011].

INDE Treat and INDE Sweet Technology : INDE Treat and INDE Sweet Technology is based on the Continuous Film contactor(CFC) for effective removal of undesirable compounds at lower cost. It can remove H_2S from LPG, Mercaptans from LPG, naphtha, gasoline and ATF/Kero, naphthenic acid from diesel, acid gases from natural gases, fuel gases and can regenerate spent caustic if required. CFC technology which is the heart of process.Salient features of CFC are

- Non-dispersive contacting

- Enormous surface are

- High mass transfer efficiency

- Based on caustic/amine

- Efficient removal of contaminants

- No aqueous phase entrainment

- Low caustic/amine consumption

- Low cost

- Can be easily retrofitted in existing mixer settler units

Merichem Fibre film Contactor Technology: The process is based on Continuous Film contactor (CFC) Fibre film Contactor technology for removal of impurities from hydrocarbon

streams.The process achieves non-dispersive phase contact without problem inherent in conventional dispersive mixing devices.

References

- Gary, J.H.; Handwerk, G.E. (1984). Petroleum Refining Technology and Economics (2nd ed.). Marcel Dekker, Inc. ISBN 0-8247-7150-8

- Fundamentals of petroleum and petrochemical engineering by Uttam Rai choudhari. Publication CRC press, International Standard Book Number: 978-1-4398-5160-9 (Hardback) chapter 3, pp. 52–53

- M. S. Vassiliou (2 March 2009). Historical Dictionary of the Petroleum Industry. Scarecrow Press. pp. 459–. ISBN 978-0-8108-6288-3

- Sadighi, S., Ahmad, A., Shirvani, M. (2011) Comparison of lumping approaches to predict the product yield in a dual bed VGO hydrocracker. , International Journal of Chemical Reactor Engineering, 9, art. no. A4

- Newton Copp; Andrew Zanella (1993). Discovery, Innovation, and Risk: Case Studies in Science and Technology. MIT Press. pp. 172–. ISBN 978-0-262-53111-5

- James H. Gary and Glenn E. Handwerk (2001). Petroleum Refining: Technology and Economics (4th ed.). CRC Press. ISBN 0-8247-0482-7

Petrochemicals: An Overview

The products obtained after petroleum distillation is called petrochemicals. The derived products include C_1, C_2, C_3 and C_4 compounds, and benzene. Petrochemicals can be classified into light petrochemicals, medium chemicals and heavy petrochemicals. The aspects elucidated in this section are of vital importance, and provide a better understanding of petrochemicals.

Petrochemicals

In this section, we present a brief overview of petrochemical technologies and discuss upon the general topology of the petrochemical process technologies.

Petrochemicals refers to all those compounds that can be derived from the petroleum refinery products

Typical feedstocks to petrochemical processes include

- C1 Compounds: Methane & Synthesis gas
- C2 Compounds: Ethylene and Acetylene
- C3 Compounds: Propylene
- C4 Compounds: Butanes and Butenes

Aromatic Compounds: Benzene

It can be seen that petrochemicals are produced from simple compounds such as methane, ethylene and acetylene but not multicomponent products such as naphtha, gas oil etc.

Definition : These are the chemicals that are made from petroleum and natural gas. Petroleum and natural gas are made up of hydrocarbon molecules, which comprises of one or more carbon atoms, to which hydrogen atoms are attached.

About 5 % of the oil and gas consumed each year is needed to make all the petrochemical products. Petrochemicals play an important role on our food, clothing, shelter and leisure. Because of low cost and easy availability, oil and natural gas are considered to be the main sources of raw materials for most petrochemicals.

Classification : Petrochemicals can be broadly classified into three categories-

a. Light Petrochemicals: These are mainly used as bottled fuel and raw materials for other organic chemicals. The lightest of these -- methane, ethane and ethylene -- are gaseous at room temperature.The next lightest fractions comprise petroleum ether and light naphtha with boiling points between 80 and 190 degrees Fahrenheit.

b. Medium Petrochemicals: Hydrocarbons with 6 – 12 carbon atoms are called "gasoline", which are mainly used as automobile fuels. Octane, with eight carbons, is a particularly good automobile fuel, and is considered to be of high quality. Kerosene contains 12 to 15 carbons and is used in aviation fuels, and also as solvents for heating and lighting.

c. Heavy Petrochemicals: These can be generally categorized as diesel oil, heating oil and lubricating oil for engines and machinery. They contain around 15 and 18 carbon atoms with boiling points between 570 and 750 degrees Fahrenheit. The heaviest fractions of all are called "bitumens" and are used to surface roads or for waterproofing. Bitumens can also be broken down into lighter hydrocarbons using a process called "cracking."

Process Topology

Reactors: Reactors are the most important units in petrochemical processes. Petrochemicals are manufactured by following simple reactions using relatively purer feedstocks. Therefore, reaction chemistry for petrochemicals manufacture is very well established from significant amount of research in this field. Essentially all petrochemical processes need to heavily depend upon chemical transformation to first product the purification.

Separation: With distillation being the most important unit operation to separate the unreacted feed and generated petrochemical product, the separation processes also play a major role in the process flow sheet. Where multiple series parallel reactions are involved, the separation process assumes a distillation sequence to separate all products from the feed. A characteristic feed recycle will be also existent in the process topology. Apart from this, other separation technologies used in petrochemical processing units include phase separators, gravity settling units and absorption columns. Therefore, the underlying physical principle behind all these separation technologies is well exploited to achieve the desired separation.

Dependence on Reaction pathway: A petrochemical can be produced in several ways from the same feedstock. This is based on the research conducted in the process chemistry. For instance, phenol can be produced using the following pathways

- Peroxidation of Cumene followed by hydrolysis of the peroxide

- Two stage oxidation of Toluene

- Chlorination of Benzene and hydrolysis of chloro-benzene

- Direct oxidation of Benzene

We can observe that in the above reaction schemes, there are two reaction pathways for phenol from benzene i.e., either chlorination of benzene or oxidation of benzene. Therefore, choosing the most appropriate technology for production is a trivial task.

Complexity in pathway: In the above Cumene example case, it is interesting to note that toluene hydrodealkylation produces benzene which can be used to produce phenol. Therefore, fundamentally toluene is required for the generation of various petrochemicals such as benzene and phenol. In other words, there is no hard and fast rule to say that a petrochemical is manufactured using a suggested route or a suggested intermediate petrochemical. Intermediate petrochemicals play a greater role in consolidating the manufacture of other downstream petrochemicals.

Synthesis gas is $H_2 + CO$ When synthesis gas is subjected to high pressure and moderate temperature conditions, it converts to methanol. Followed by this, the methanol is separated using a series of phase separators and distillation columns. The process technology is relatively simple.

Reactions

- Desired: $CO + 2H_2 \rightarrow CH_3OH$

- Side reactions: $CO + 3H_2 \rightarrow CH_4 + H_2O$

$$2CO + 2H_2 \rightarrow CH_4 + CO_2$$

All above reactions are exothermic

- Undesired reaction: zCO + aH2→ alchohols + hydrocarbons.

- Catalyst: Mixed catalyst made of oxides of Zn, Cr, Mn, Al.

Process Technology

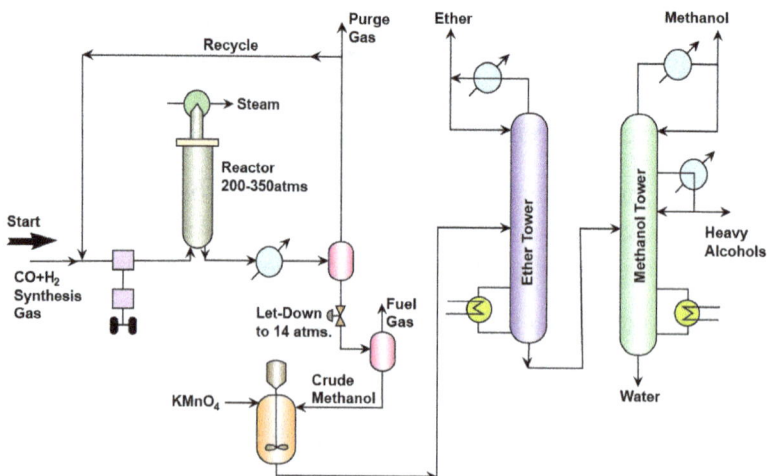

Flow sheet of manufacture of Methanol from Synthesis Gas

- H_2 and CO adjusted to molar ratio of 2.25.

- The mixture is compressed to 200 – 350 atms.

- Recycle gas (Unreacted feed) is also mixed and sent to the compressor.

- Then eventually the mixture is fed to a reactor. Steam is circulated in the heating tubes to maintain a temerature of 300 – 375°C.

- After reaction, the exit gases are cooled.

- After cooling, phase separation is allowed. In this phase separation operation methanol and other high molecular weight compounds enter the liquid phase and unreacted feed is produced as the gas phase.

- The gas phase stream is purged to remove inert components and most of the gas stream is sent as a recycle to the reactor.

- The liquid stream is further depressurized to about 14 atms to enter a second phase separator that produces fuel gas as the gaseous product and the liquid stream bereft of the fuel gas components is rich of the methanol component.

- The liquid stream then enters a mixer fed with $KMNO_4$ so as to remove traces of impurities such as ketones, aldehydes etc.

- Eventually, the liquid stream enters a distillation column that separates dimethyl ether as a top product.

- The bottom product from the first distillation column enters a fractionator that produces methanol, other high molecular weight alcohols and water as three different products.

Structure of Petrochemical Complexes

The petrochemical complexes involve one or a combination of the following operations :

- The manufacture of basic raw materials like syngas, methane, ethylene, propylene, acetylene, butadiene, benzene, toluene, xylenes, etc. The basic building processes include partial oxidation, steam reforming, catalytic and thermal cracking, alkylation, dealkylation, hydrogenation, disproportionation, isomerisation, etc. The commonly used unit operations are distillation, extractive distillation, azeotropic distillation, crystallisation, membrane separation, adsorption, absorption, solvent extraction, etc.

- Manufacture of intermediate chemicals derived from the above basic chemicals by various unit processes like oxidation, hydrogenation, chlorination, nitration, alkylation, dehydrogenation along with various unit operations like distillation, absorption, extraction, adsorption, etc.

- Manufacture of target chemicals and polymers that may be used in the manufacture of target products and chemicals to meet the consumer needs. It includes plastics, synthetic fibers, synthetic rubber, detergents, explosives, dyes, intermediates, and pesticides.

First Generation Intermediates	Hydrogen, Ammonia, Methanol Olefins and Dienic Hydrocarbons, Ethylene, Propylene, Butadiene, Isoprene, etc. Aromatic Hydrocarbons, Benzene, Toluene, Xylenes, Styrene, etc.
Second Generation Intermediates	Introduction of various hetro atoms into final molecule including oxygen, nitrogen, chlorine and sulfur by various unit process Intermediates
Target Product	Plastics, Synthetic fibre, Fertilizers, Solvents, elastomer, Drugs, Dye stuff, Detergent, Explosive, Pesticides.

Indian Petrochemical Industry

Petrochemicals are backbone of chemical industry. Although the origin of petrochemical industry in world was in 1920, however, sixties marks the era of petrochemicals in India when Union Carbide set up first ethylene complex with capacity 20,000 TPA in 1963 in Mumbai. It was followed by NOCIL with 60,000 TPA ethylene complexes in 1968 in Thane near Mumbai and PSF plant of Chemical and Fibers India Ltd.(CAFI) at Thane - Belapur road. Indian Petrochemical Corp. Ltd. (IPCL) set up first integrated petrochemical complex in 1970 in public sector at Vadodara. With the modest beginning in 1960 by setting up of 20,000 tonnes naphtha cracked by Union Carbide in Mumbai, Indian petrochemical industry has sustained a high and steady rate of growth during last four decades and has entered the world market. The petrochemical industry has been the fastest growing sectors in India and become a major segment of chemical industry, which is growing faster than industries overall and within chemicals. It has posed serious threat to chemical industry based on natural feedstock – Biomass and Coal. The petrochemical industry is major supplier of chemical inputs to a large and growing number of downstream. The Indian chemical industry ranks twelfth by volume in the world for production of chemicals and accounts for about 14percent in the general index of industrial production and 17.6percent in the manufacturing sector. It accounts for 13-14percent of total exports and 8-9percent of total imports of the country. It contributes about 3 percent to the GDP [Annual Report, 2004-05]. It contributes 13percent of the manufacturing industry's value added and 8percent of the total exports of the country.

Basic Petrochemicals

C1 group   Methane, CO – H_2 synthesis, synthesis gas derivatives

C2 group   Ethane, ethylene, ethylene derivatives, acetylene

C3 group   Propane, propylene and propylene derivatives

C4, C5 group   Butadiene, Butanes, Butenes, Pentane, Pentene, Isoprene, Cyclopentadiene

Aromatic   Benzene, Toluene , Xylenes Naphthalene, BTX derivatives

Major End Products

Polymer, Synthetic fibre, synthetic rubber, synthetic detergent, Chemical intermediate, dyes and intermediates chemical intermediates, pesticides

Basic Building Block Process

Petrochemical manufacturing involves building blocks processes for the manufacture of building blocks and intermediates.

Cracking: Steam cracking, Catalytic cracking for olefins pyrolysis gasoline by product

Steam reforming and Partial oxidation: Synthesis gas

Catalytic Reforming: Aromatic production

Aromatic conversion processes: Aromatic production

Alkylation: Linear alkyl benzene

OXO Process: Oxo-alcohol

Polymerisation Process: Polymer, elastomers and synthetic fibre

Petrochemicals Feed Stock

One of the major issues, which have directed the worldwide growth of petrochemical industry, has been the availability of feedstock, which has led to replacement of the natural resources coal, molasses, fats, etc. Basic feedstock used in petrochemical industry for manufacture of olefins and aromatics are derived either from natural gas or petroleum fractions and includes natural gas (associated or non-associated), condensate, naphtha, kerosene, catalytic cracking and reformer gases, waxes, pyrolysis gasoline. Natural Gas and Petroleum Fractions as Petrochemicals Feedstock is given in Table. Alternative routes to principal petrochemicals is given in Table. Some of the alternative feed stock choice for petrochemical industry are:

- Naphtha from methane from natural gas to liquid process

- Naphtha from coal via direct liquification or indirect liquification by FT process

- Plastic waste to naphtha and other hydrocarbons through liquefaction, pyrolysis and separation processes

- FT naphtha from biomass

- Methanol routes: Synthesis gas from methane, coal and biomass; conversion of synthesis gas to methanol and production of olefin by methnol to olefin technology.

- Conversion of methanol to dimethyl ether

- Product recovery and separation Recovery of C_4 & C_5 stream from FCC and steam cracker

- Oxidative coupling of methane

- Ethanol from biomass: direct fermentation of sugar rich biomass, hydrolysis of lingo- cellulosic biomass

- Gasification of lingo-cellulosic biomass followed by fermentation or chemical catalysis to ethanol.

- Carbon dioxide to liquid fuel by engineered bacteria

- Gasification of petrocoke to hydrogen

Table: Natural Gas and Petroleum Fractions as Petrochemicals Feedstock

Petroleum Fractions and Natural Gases	Source	Composition	Intermediate Processes	Intermediate Feedstock
Refinery **Gases**	Distillation, cata-lytic cracking, cat-alytic reforming	Methane, ethane, propane, butane, BP upto 25 $^{\circ}C$	Liquefaction, cracking	LPG, ethylene propylene, butane, butadiene.
Naphtha	Distillation and thermal & cat-alytic cracking, visbreaking	C_4-C_{12} hydrocar-bon, BP 70 - 200 $^{\circ}C$	Cracking, reform-ing, alkylation, disproportionation , isomerisation	Ethylene, propyl-ene, butane, bu-tadiene, benzene, toluene, xylene
Kerosene	Distillation and secondary conver-sion processes	C_9-C_{10} hydrocar-bon, BP 175-275 $^{\circ}C$	Fractionation to obtain C_{10}-C_{14} range hydrocarbon	Linear n C_{10} - n C_{14} alkanes
Gas Oil	Distillation of crude oil and cracking	C_{10}-C_{25} hydrocar-bons BP 200-400 $^{\circ}C$	Cracking	Ethylene, propyl-ene, butadiene, butylenes
Wax	Dewaxing of lubri-cating oil	C_8-C_{56} hydrocarbon	Cracking	C_6-C_{20} alkanes
Pyrolysis Gasoline	Ethylene cracker	Aromatic, alkenes, dienes, alkanes, cycloalkane	Hydrogenation distillation, ex-traction, crystalli-sation, adsorption	Aromatics
Natural Gases & Natural Gas Condensate	Gas fields and crude oil stabili-sation	Hydrogen, methane, ethane, propane, pentane, aromatics	Cracking, reform-ing, separation	Ethylene, propyl-ene, LPG, aromat-ics, etc.
Petroleum coke	Crude oil	Carbon	Residue upgra-dation processes, gasification	Carbonelectrode, acetylene, fuel

Table: Alternative Routes to Principal Petrochemicals

Chemicals	Petroleum Source	Alternate Source (Europe, except where stated)
Methane	Natural gas Refinery light gases (demethaniser overheads)	Coal, as byproduct of separation of coke oven gases (1920-30) or of coal hydrogenation (1930-40)
Ammonia	Methane Light liquid hydrocarbons	From coal via water gas (1910-20)
Methyl alcohol	Methane Light liquid hydrocarbons	From coal via water-gas (1920-30); from methane (from coal) by methane-stream and methane oxygen processes (1930-40)
Ethylene	Pyrolysis of gaseous liquid hydro-carbons	Dehydration of ethyl alcohol (original route). By-product in fractional distillation of coke oven gas (1925-35). Hydrogenation of acetylene (1940-45)
Acetylene	Ethylene	Calcium carbide (original process). methane from coal by partial combustion and by arc process (1935-45)
Ethylene glycol	Ethylene	From ethylene made as above (1925). In America, from coal via carbon-monoxide and formaldehyde (1935-40)
Ethyl alcohol	Synthetic ethyl alcohol Co-product	Fermentation of molasses (original route)
Acetaldehyde	of paraffin gas oxidation. Direct oxidation of ethylene	Fermentation of ethyl alcohol, or acetylene from carbide (1900-10)
Acetone	Propylene	Wood distillation (original process). Pyrolysis of acetic acid (1920-30) or by acetylene-stream reaction (1930-40)

Glycerol	Propylene	By-product of soap manufacture (original process)
Butadiene	1- and 2-Butenes Butane Synthetic ethyl alcohol By-product of ethylene by py-rolysis of liquid hydrocarbons	Ethyl alcohol (1915); acetaldehyde via 1:3- butanedi-ol (1920-30); acetylene and formaldehyde from coal via 1:4-butanediol (1940-45); from 2:3-Butanediol by fermentation (1940-45)
Aromatic hy-drocarbons	Aromatic-rich and naphthen-ic-rich fractions by catalytic reforming and direct ex-traction or by hydroalkylation	By-products of coal-tar distillation

Evaluation of Feed Stocks For Aromatics, Olefins and Surfactant Plants

AROMATICS   -Naphtha, Pyrolysis gasoline, LPG

OLEFINS   -Ethane, Propane, Naphtha, Gas oil

SURFACTAN PLANTS   -Kerosene for paraffins, benzene

- Input cost of feed constituents is a major portion of the variable cost of production in petrochemical plants.

- Major feed input naphtha/kerosene from the refinery

- Fed quality monitoring and improvement efforts are therefore very important aspects having significant impact on the economics of the operation cost.

- The precursors and undesirable constituents in feed including catalyst and adsorbents poisons should be known, analyzed and monitored continuously.

Desired Components in Feed for Olefins Productions

- Naphthenes: Naphthene yield olefins of higher carbon number. Butane yield increases appreciable with naphthenic feed. Naphthenes also enhance production of aromatics.

- Aromatics: The aromatics is feed are highly refractory and they pass through the furnace unreacted.

- Sulphur: The sulphur in feed suppresses stream reforming reaction catalyzed by nickel present in radiant coil. Optimum level of sulphur- 1 ppm.

Physical Properties: Density, distillation range are useful and give a rough assessment of feed quality.

Ethylene

The following components in feed give ethylene in decreasing order:

Ethane, Butane to Decane, 3 and 2 Methyl hexane, 2 methyl Pentane/ 2,2 Dimethyl Butane, Isopentane

Propylene

The following components in feed give propylene in decreasing order:

Isobutane, n-nutane, n-propane, 3 methyl pentane, 2,3 dimethyl butane, 2 methyl hexane, n-pentane, 3 methyl hexane, iso pentane.

Butadiene

The following components of feed give butadiene is decreasing order: Cyclo hexane, methyl cyclo pentane.

Aromatic Plant

Naphtha cut C_6 to C_9

Paraffin, Napthenes, Aromatics 110 to $140^\circ C$

Dehydrogenation of C_8 Napthene yield C_8 aromatics. Most desirable component 90% of C8napthalene in feed gets converted to C_8 aromatics

- C_8 Paraffin's: Dehydro cyclisation of C_8 paraffin's yield aromatics difficult to 20% C_8 paraffins converted to C_8 aromatics.

- C_8 aromatics: Pass as refractory and directly contribute to   aromatic production.

- C_8 aromatic precursors: It is useful to monitor aromatic precursors= 0.2* C_8 P + 0.9 * C_8 N + 1.0 C_8 A.

Integration of Refinery with Petrochemical

Advances in processing technologies are playing a larger role in integrating refining and petrochemical facilities. In the changing scenario, petroleum refining and petrochemical production integration will be of vital importance for maximizing the use of byproducts and improving the overall economy of a petroleum refinery. A great deal of synergy exists between the refinery, aromatics complexes and steam cracker complex. Off gases from the FCC unit and coker containing ethylene and propylene can be integrated with the cold section of steam cracker. Pyrolysis gasoline is a good source of aromatics which can be integrated with the catalytic reforming process. Propylene from FCC and benzene from aromatics are feed stocks for the production of cumene and phenol. A new concept of refinery petrochemical integration are:

- Low to moderate level of integration: uses 5-10 % of crude

- High level integration: these complexes convert 10-25% of the crude oil

- Petrochemical refinery: these complexes produce a significant amount of petrochemicals as compared to fuels.

Petrochemical Processes within refinery which will help in integration of refinery and petrochemicals

- Propylene Recovery from FCC gases

- Ethylene from FCC gases

- C_4 and C_5 recovery from FCC

- $C_4 s$ from naphtha cracker and refinery to LPG pool as well as feed to cracker

- Aromatic Recovery & Conversations

- Light ends & Light Naphtha Conversion

- Residue & Coke gasification

- Hydrogen Production

- Butane to Maleic anhydride & Derivative

- Benzene-Cumene-Phenol-Acetone

- Benzene-Cyclohexane-Caprolactum

- n-Paraffins extraction from kerosene for LAB

- Valorization of refinery streams- LCO, LCGO, HCGO

- Recovery of Valuable Chemicals cyclopentadiene, diclopentadiene, isoprene, propolyene

- Isobutylene for alkylation

- Use of C7-C8 stream from benzene extortion for separation of p-xylene for PTA

- Maximizing the use of natural gas in a refinery-petrochemicals complex offers higher margins and lower carbon emissions. Off gas from Off gas from FCC and delayed coking units contains a good quantity of ethane, ethylene, propylene and some propane and recovery of these hydrocarbons may be economical in the refinery.

Indalin Process

Indalin is a versatile indigenous technology adding value to upstream and downstream oil industry. Indalin is a catalytic cracking process for upgradation of low value naphtha to very high yield of LPG, containing high olefins such as propylene, ethylene butylenes etc. Surplus kerosene and gas oil range fraction can also be processed along with naphtha. Indalin can integrate a refinery with petrochemicals complex and therefore offers a tremendous opportunity for value addition through upgradation of low value streams to petrochemical feed stock.

Hydrocarbon Steam Cracking

With the rising demand of ethylene and propylene, there has been a tremendous growth in the steam cracking of hydrocarbons during the last four decades. Similarly, FCC (Fluid Catalytic Cracking) has developed into a major upgrading process in the petroleum refinery industry for the conversion of heavy fuel oil into more valuable products ranging from light olefins to naphtha and middle distillate. Large amounts of C_4 and C_5 compounds are produced along with the production of ethylene in steam cracking and gasoline in FCC. C_4 & C_5 streams are an important source of feedstock for synthetic rubber and many chemicals.

With increasing demand of C_5 hydrocarbons and oxygenates, upgrading of C_4 and C_5 streams from steam crackers and catalytic cracker is important to the economic performance of the above processes. It also provides a rich resource of reactive molecules, which forms the backbone of the synthetic rubber industry. The quantity and composition of the C_4 and C_5 stream depends on the severity of the steam cracker operation and feedstock processed.

Product profile C4 and C5 hydrocarbons are given in Figure and Table.

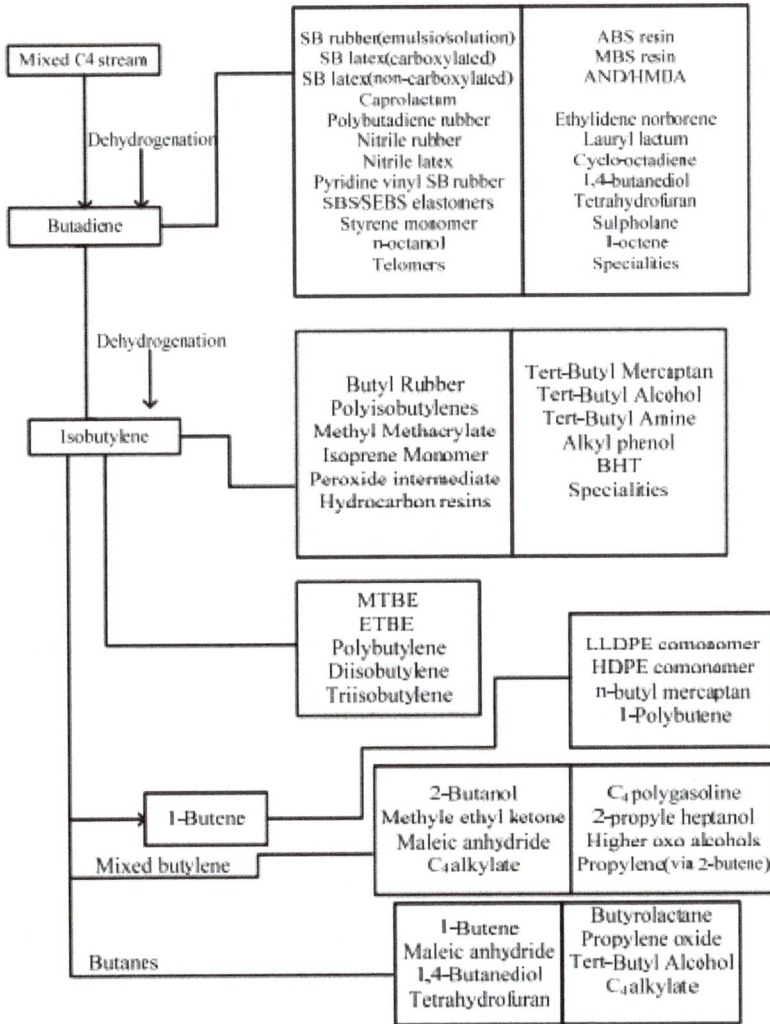

Product Profile of C4 and C5 Hydrocarbon

Product Profile of C5 Hydrocarbon

C_5 hydrocarbons –are an important source of synthetic rubber, solvents, chemical intermediate, MTBE, plasticisers, TAME, rubber chemicals, herbicides, lube oil additives, pharmaceuticals.

Table: Product Profile C$_5$ Hydrocarbons

C$_5$ Hydrocarbon	Isoprene	Polyisoprene, as the cross linking agent in Butyl rubber As co-monomer in stryrene-isoprene copolymers
	Isopentane	Solvent, Chlorinated derivative, blowing agent for Polystyrene
	1-Pentene	Organic Synthesis, blending agent for high octane fuel
	2-Pentene	Polymerisation inhibitor, organic synthesis
	CycloPentene	Organic synthesis, polyolefins,epoxies cross linking agent
	2- Methyl-1-Butene	Synthetic mark, anyl benzene hydrogen synthetic mark, anyl benzene hydrogen peroxide catalyst, 2,4-diamyl phenol (photographs colour complex), pinacolone (Crop protection chemicals)
	3- Methyl-1-Butene	Monomer for specialty homo-polymer
	Cyclopentadiene	Chlorinated insecticides, Chemical intermediate, Antiviral-agent
	Piperylene	Polymers, maleic anhydride,chemical intermediate

Fluid Catalytic Cracking

Fluid catalytic cracking (FCC) converts low value crude oil into a variety of higher value products which include gasoline, diesel, heating oil and valuable gases containing LPG, propylene and C4 and C5 gases. Various products from fluid catalytic cracking and their uses are given in Table. FCC units are versatile and can be operated in three main modes which are aimed at maximizing middle distillate, gasoline, or olefins respectively by means of the adequate combination of various parameters such as catalyst type, catalyst to oil ratio, rise of outlet temperature and recycle of fractionators bottom. FCC is the second largest source of propylene supplied for petrochemical application.

- Conventional FCC 4-7% propylene and 1-2 % Ethylene

- High Severity FCC:10% propylene

- Petro FCCTM (UOP): Ethylene 6%, Propylene 20-22%, Higher aromatics (18%) in Naphtha

- Higher C_{4-8} olefins yield which can be cracked to yield lower olefins by Total Petrochemicals ATOFINA/UOP Olefin cracking Process

- Although FCC is an important petroleum refining process, however, FCC gases have now become important petrochemical feedstock for production of LPG that can be converted to aromatics and C3, C4, & C5 hydrocarbons, i.e. propylene, butene, isobutene, pentene, etc.

Product distribution from FCC depends

- Reactor temp

- Feed preheat temperature

- Catalyst activity

- Catalyst circulation rate

- Catalyst activity

- Recycle rate

Table: Various Petroleum Products from FCC and their uses

Product	Composition and Uses
Light gases	Primarily H_2, C_1 and C_2s, ethylene can be recovered
LPG	C_3s and C_4s containing light olefins suitable for alkylations
Gasoline	C_5+ high octane component for gasoline pool or light fuel
Light cycle oil (LCO)	Blend component for diesel or light fuel
Heavy cycle oil (HCO)	Fuel oil or cutter oil
Clarified oil	Carbon black feedstock
Coke	Used in regenerator to provide the reactor heat demand

Propylene Recovery from FCC: FCC gases has important source of propylene from refinery and now FCC units are being operated both in gasoline mode and propylene mode. Propylene from FCC may be as high as 25% with new FCC based propylene technologies. increased production of olefins from FCC unitsc has been achieved through changes in operations,base cracking catalyst and additive catalysts . and in hardware designs.

Upgrading of C_4 and C_5 Streams

C_4 and C_5 Streams from Steam Cracker and FCC contains C4 and C5 hydrocarbons recovery of which has become important steps for improving the overall economy of these processes. Some of the important C4 streams from Cracker and FCC butadiene(-from cracker plant only) ,butene- 1, 2- butane, isobutylene, mixed n-butene, isobutene.. C4 stream of steam cracker contains appreciable amount of butadiene which is being recovered from naphtha cracker plants. Typical composition of C4 stream of naphtha cracker and FCC is given Table. The distribution product will depend on thefeed stock,cracking severity and catalyst in case of FCC

Table: Typical Compositions of C4 Fractions

Component	FCC	Steam Cracking
Isobutane	37.0	2.0
Isobutene	24.0	26.0
1-Butene	15.0	13.6
1,3-Butadiene	0.2	36.0
2-Butenes (cis and trans)	11.0	12.0
n-Butane	12.0	9.8
others	balance	balance

Typical C_5 cuts from steam cracking contain C_4 (1%), n-pentene (26%), isopentane (24%), n- pentenes (4.5%), methyl butenes (12%), cyclopentenes (1.5%), isoprene (13.5%), pentadiene (piperylene) (9.0%), cyclopentadiene (7.5%), C_6+ (1%) [Chauvel& Lefebvre, 1989]. Cyclopentadiene is easily dimerised to higher boiling dicyclopenta-diene and separated from C_5 stream by simple distillation. Typical composition of C_5 cuts from catalytic cracking may be C_4(2%), n-pentane (5.5%), isopentane (31.5%), n-pentenes (22.5%), methyl butenes (37.5%), C_6 + (1%) [Chauvel & Lefebvre, 1989]. Naphtha feed gives higher yield of C_4 (8-10%) than ethane feed (2-3%)

- Upgrading of C_4 Olefins :

- The production of chemical intermediates

- Butene-1, isobutylene, mixed n-butene

- Production of motor fuel component (alkylate, dimate, MTBE)

Processing of C_4 cut from Steam Cracker and FCC

There is not much difference in the processing of C_4 streams after the recovery of bu-tadiene from the steam cracker and C_4 streams from the FCC. C4 stream Butadiene from C_4 stream of naphtha cracker/ gas cracker is first recovered, followed by sepa-ration Isobutylene, isobutanee, butane, butane-1 and butene-2 from C4 stream/ FCC and cracker using various process like etherification, hydrolusis, cracking, adsorption distillation etc. plant by various

C_4 cut from steam cracker and FCC is shown in Figure. Isobutene recovery includes either hydration of the C_4 stream and subsequent decomposition or etherification with methanol to yield MTBE, which is cracked to give isobutene. Separation of 1-butene is done by selective hydrogenation followed by adsorption for separation of 1-butene and further processing for separation of isobutene and 2-butene by distillation. Separation of 2-butene involves hydro- isomerisation and subsequent distillation for separation of isobutene and 2-butene.

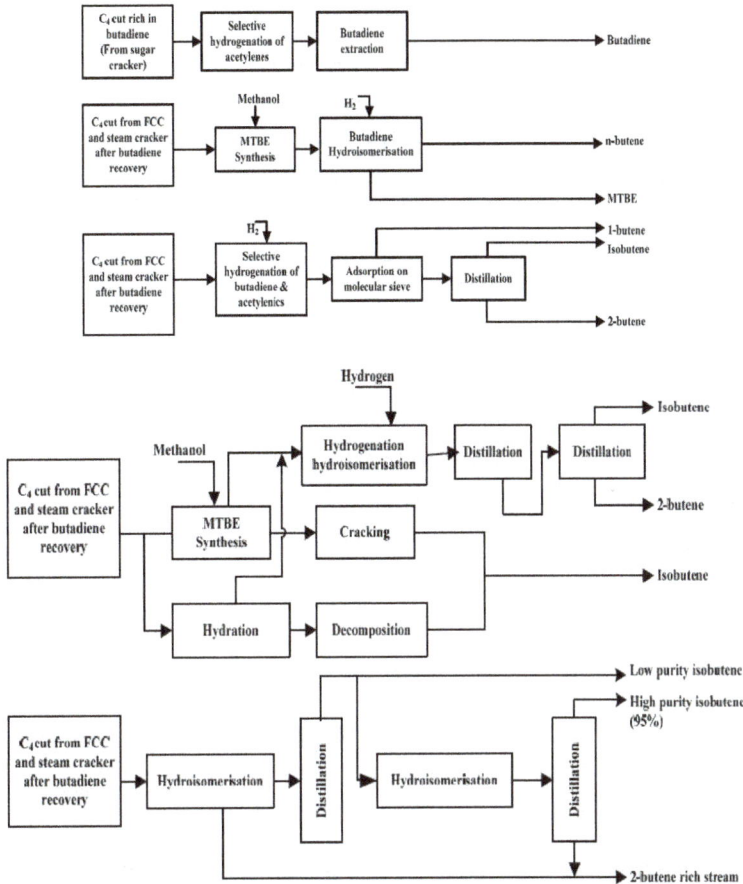

Separation of C4 hydrocarbons from FCC and Steam Cracker plants

After separation of butadiene, the C_4 streams from cracking and FCC is processed for production of n-butene, 1-butene, 2-butene, and isobutene. Process flow diagram for treatment of

Butadiene

Butadiene is important raw material for production of a larger number of synthetic rubber and polymers such as styrene butadiene rubber (SBR), poly butadiene, chloroprene rubber, nitrile rubber, acrylonitrile butadiene styrene plastic. Other fastest growing use is in the manufacture of adiponitrile used in the manufacture of Nylon 66, Steam cracker and catalytic dehydrogenation of butenees are the two major sources of butadiene. Butenes can be recovered from C_4 stream or produced by dehydrogenation of butanes.

According to SRI consulting global production and consumption of butadiene in 2009 was approximately 9.2 million tones and 9.3 million tones respectively. Butadiene is expected to average growth of 4.9 per year from 2009-2014. Styrene –butadiene rubber accounted for more than 33% of global butadiene consumption in 2009 and butadiene rubber for about 25%.

Polymerisation Grade Butadiene 1,3 Butadiene % min	99.6%
Butenes ppm max	4000
Methyacetylenes ppm.max.	25
Vinyl acetate ppm max	200
C5 dimers ppm max	2000
Carbonyl compounds (as aldehyde) max.	50
Inhibitor (p-tertiobutyl catechol)	100- 200
Non-volatile residue ppm	2000

There are four major routes for production of butadiene:

- Steam cracking of naphtha

- Catalytic dehydrogenation of butenes

- Catalytic dehydrogenation of butanes

- Dehydrogenation-dehydration of ethanol (molasses route)

Butadiene from C_4 stream of Cracker Plant

C_4 cut from the steam cracker is first sent for butadiene recovery, which includes selective hydrogenation of acetylenics in the presence of palladium catalyst, then separation of butadiene extractive distillation process steps involved are

- Extractive distillation in which acetylnic compound and butadiene are extracted in one or two stages

- Recovery of solvent

- Super fraction of butadiene stream for removal of acetylnic impurities

- Water scrubbing butadiene depleted cut to recover the solvent.

Various solvents used for separation of butadiene are furfural, dimethyl formamide (DMF), n- methyl pyrrolidone (NMP), and dimethyl acetamide. Selective hydrogenation results in overall improvement in the economy with higher butadiene yield.

Catalytic Dehydrogenation of Butenes

Reaction:

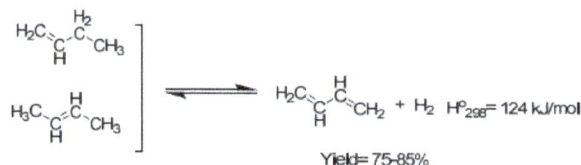

Yield= 75-85%

Catalytic dehydrogenation of butanes two stages process:

Catalytic dehydrogenation of butanes to butenes

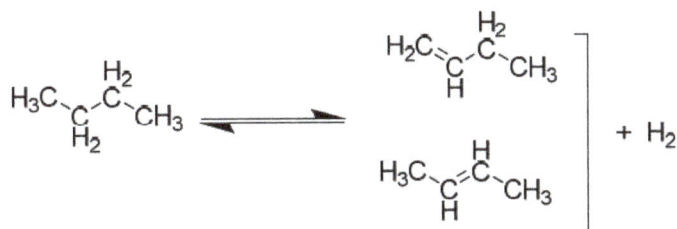

Catalytic dehydrogenation of butanes to butadience

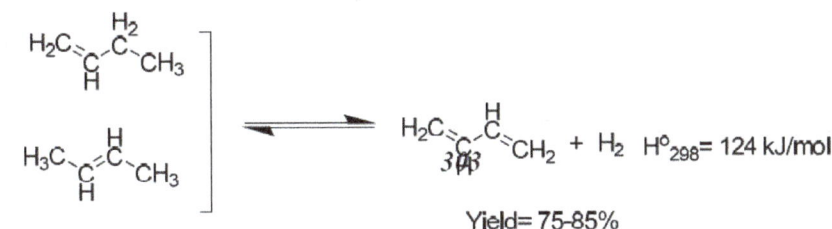

$$H^\circ_{298} = 124 \text{ kJ/mol}$$

Yield= 75-85%

Isobutylene

Isobutylene is present in the C_4 stream naphtha cracker and FCC. Major application of isobutene is in the manufacture of gasoline blending component such as MTBE, ETBE, alkylation, polymer gasoline. Polymer grade isobutylene can be made by cracking MTBE or for manufacture of polyisobutylene. Isobutylene is used in manufacture butyl rubber which is made by copolymerization of isobutylene with small amount of isoprene.

Various Routes for Isobutylene

Extraction of C_4 cuts from steam cracking / FCC: Isobutylene is separated from C_4 cuts from naphtha cracker after extraction of butadiene and from FCC gases after propylene recovery. First isobutylenes is converted to MTBE by etherification and the recovered by cracking of MTBE to get polymer grade isobutylene it is also obtained by hydration of isobutylene containing stream and then cracking.

Isomerisation of Butene: isobutylene can be also produced from butane by isomerisation using zeolite ferrierite (zeolite of medium pore size) [Maulijanet al.2001]

Dehydrogenation of Isobutene:

BUTENE -1

Butene-1 is co-monomer in the production of low density polyethylene and high density polyethylene. Butene-1 can be separated from C4 stream of cracker after extraction of butadiene SHB-CB process: This process selectively hydrogenate the butadiene in the C4 cut by converting it to butane-1 and butane-2. Acetylenes and dienes are likewise hydrogenated. If the process is optimized to produce butane-1, about 60% of butadiene is converted to butane-1. The process is operated in the liquid phase mild temperatures and moderate pressures.

Upgrading of C_5 Cuts

The steam cracker C5 stream is a rich resource of olefins and diolefins which can be upgraded to produce elastomers, resins and fine chemical intermediates. In steam crackers during cracking process along with ethylene, propylene, C4 stream, aromatics and pyrolysis gasoline bare also formed. Apart from aromatics, Pyrolysis gasoline stream also contains C5 stream. The quantity and composition of the stream depend on the nature of the cracked product and severity of cracker operation C_5 stream.

Various Steps in the recovery of C_5 chemicals are:

- Separation of C_5 stream from pyrolysis gasoline by distillation

- Separation of cyclopentadiene: In first stage cycolpentadine is dimerised to dicyclopentadiene followed by cracking of dicylopentadiene to cyclopentadiene.

- Extractive distillation of cyclopentadiene free C_5 stream produce isoprene-piperylene stream. Distillation removes the light aceyelenes

- Separation of isoprene and piperylene extract by distillation

- Absorption at atmospheric pressure in the presence of NMP

- Purification of Isoprene rich paraffin

- Periodic regeneration of solvent

Solvents used in extraction of isoprene are Acetonitrile, N-methylpyropedone, Dimethylformalnide

Oxygenates from Refinery C_4 and C_5 Stream

Several oxygenated fuel components have figured prominently in refinery reformulated gasoline planning. Methyl tertiary butyl ether (MTBE), tertiary amyl methyl ether (TAME) and ethyl tertiary butyl ether (ETBE). All oxygenated fuels reduce hydrocarbons in the automobile exhaust. MTBE was considered one of the most important oxygenates used in the production of lead free gasoline and was used produced on a large scale throughout the world. There has been because of environmental problem. The oxygenated MTBE and ETBE are produced by the reaction of methanol/ethanol and isobutylene.

Methyl Tertiary Butyl Ether (MTBE)

MTBE is one of the important oxygenates and originally its use started as a substitute of tetraethyl lead. MTBE increases the oxygen content of gasoline results in the reduction of harmful emissions. MTBE which is made by etherification of C4 gases from cracker and FCC is also used for production of polymer grade isobutylene for synthetic rubber.

MTBE is produced by the reaction of methanol with isobutylene contained in C_4 streams from thermal crackers in the presence of ion exchange resin at $40-90^{\circ}C$ and a pressure of 5 to 10 $kg/cm^2 g$. Catalytic cracking butylenes and field butanes are additional possible source of isobutylene. Convention process and catalytic distillation are the two commercial processes

available. Figure shows the process flow diagram fro MTBE conventional methods.

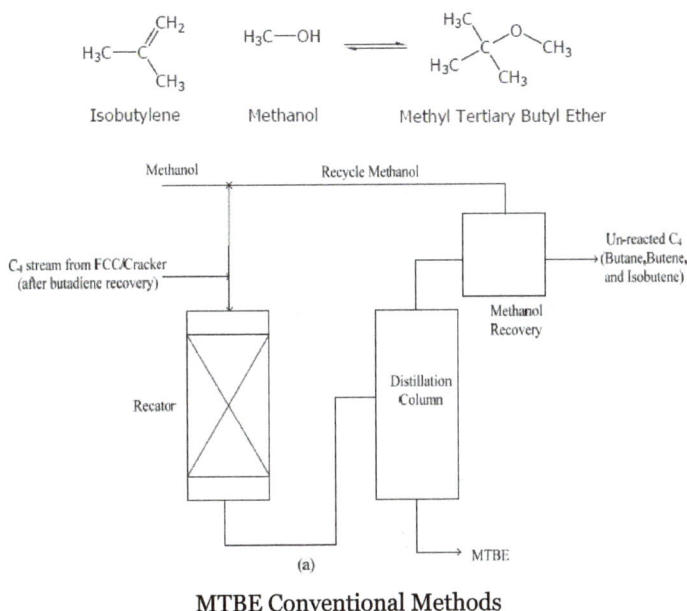

MTBE Conventional Methods

Ethyl Tertiary Butyl Ether (Etbe)

ETBE is made by etherification of isobutylene with ethanol similar to MTBE.

Isobutylene + Ethanol → ETBE

Teriary Amyl Methyl Ethertame

TAME is produced by etherification of isoamylenes recovered from C_5 stream of FCC and steam crackers. Two reactive components of iosmaylenes are 2-Methyl butene-1 and 2-Methyl butene- 2. Catalytic distillation process is used for the manufacture of TAME.

Reaction Involve:

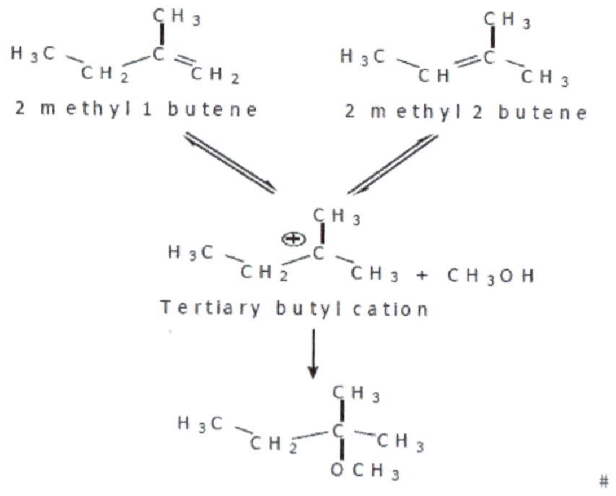

Syngas

Syngas, or synthesis gas, is a fuel gas mixture consisting primarily of hydrogen, carbon monoxide, and very often some carbon dioxide. The name comes from its use as intermediates in creating synthetic natural gas (SNG) and for producing ammonia or methanol. Syngas is usually a product of gasification and the main application is electricity generation. Syngas is combustible and often used as a fuel of internal combustion engines. It has less than half the energy density of natural gas.

Syngas can be produced from many sources, including natural gas, coal, biomass, or virtually any hydrocarbon feedstock, by reaction with steam (steam reforming), carbon dioxide (dry reforming) or oxygen (partial oxidation). Syngas is a crucial intermediate resource for production of hydrogen, ammonia, methanol, and synthetic hydrocarbon fuels. Syngas is also used as an intermediate in producing synthetic petroleum for use as a fuel or lubricant via the Fischer–Tropsch process and previously the Mobil methanol to gasoline process.

Production methods include steam reforming of natural gas or liquid hydrocarbons to produce hydrogen, the gasification of coal, biomass, and in some types of waste-to-energy gasification facilities.

Production Chemistry

The chemical composition of syngas varies based on the raw materials and the processes. Syngas produced by coal gasification generally is a mixture of 30 to 60% carbon monoxide, 25 to 30% hydrogen, 5 to 15% carbon dioxide, and 0 to 5% methane. It also contains lesser amount of other gases.

The main reaction that produces syngas, steam reforming, is an endothermic reaction with 206 kJ/mol methane needed for conversion.

The first reaction, between incandescent coke and steam, is strongly endothermic, producing carbon monoxide (CO), and hydrogen H_2 (water gas in older terminology). When the coke bed has cooled to a temperature at which the endothermic reaction can no longer proceed, the steam is then replaced by a blast of air.

The second and third reactions then take place, producing an exothermic reaction—forming initially carbon dioxide and raising the temperature of the coke bed—followed by the second endothermic reaction, in which the latter is converted to carbon monoxide, CO. The overall reaction is exothermic, forming "producer gas" (older terminology). Steam can then be re-injected, then air etc., to give an endless series of cycles until the coke is finally consumed. Producer gas has a much lower energy value, relative to water gas, due primarily to dilution with atmospheric nitrogen. Pure oxygen can be substituted for air to avoid the dilution effect, producing gas of much higher calorific value.

When used as an intermediate in the large-scale, industrial synthesis of hydrogen (principally used in the production of ammonia), it is also produced from natural gas (via the steam reforming reaction) as follows:

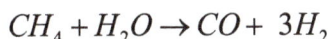

$$CH_4 + H_2O \rightarrow CO + 3H_2$$

In order to produce more hydrogen from this mixture, more steam is added and the water gas shift reaction is carried out:

$$CO + H_2O \rightarrow CO_2 + H_2$$

The hydrogen must be separated from the CO_2 to be able to use it. This is primarily done by pressure swing adsorption (PSA), amine scrubbing, and membrane reactors.

Alternative Technologies

Biomass Catalytic Partial Oxidation

Conversion of biomass to syngas is typically low-yield. The University of Minnesota developed a metal catalyst that reduces the biomass reaction time by up to a factor of 100. The catalyst can be operated at atmospheric pressure and reduces char. The entire process is autothermic and therefore heating is not required.

Carbon Dioxide and Hydrogen

Microwave Energy

CO_2 can be split into CO and then combined with hydrogen to form syngas . A method for production of carbon monoxide from carbon dioxide by treating it with microwave radiation is being examined by the solar fuels-project of the Dutch Institute For Fundamental Energy Research. This technique was alleged to have been used during the Cold war in Russian nuclear submarines to allow them to get rid of CO_2 gas without leaving a bubble trail. Publicly available journals published during the Cold War indicate that American submarines used conventional chemical scrubbers to remove CO_2. Documents released after the sinking of the Kursk, a Cold War era Oscar-class submarine, indicate that potassium superoxide scrubbers were used to remove carbon dioxide on that vessel.

Solar Energy

Heat generated by concentrated solar power may be used to drive thermochemical reactions to split carbon dioxide to carbon monoxide or to make hydrogen. Natural gas may be used as a feedstock in a facility that integrates concentrated solar power with a power plant fueled by natural gas augmented by syngas while the sun is shining. The Sunshine-to-Petrol project has developed a device allowing for efficient production using this technique. It is called the Counter-Rotating Ring Receiver Reactor Recuperator, or CR5.

Wind Energy

An airborne wind energy system has been proposed to supply heat to the steam reforming reaction. This avoids burning natural gas for the heat and radically simplifies the steam reformer.

Co-Electrolysis

By employing co-electrolysis, i.e. the electrochemical conversion of steam and carbon dioxide with the use of renewably generated electricity, syngas can be produced in the framework of a CO_2-valorization scenario, allowing for a closed carbon cycle.

Electricity

Use of electricity to extract carbon dioxide from water and then water gas shift to syngas has been trialled by the US Naval Research Lab. This process becomes cost effective if the price of electricity is below $20/MWh.

Renewable Sources

Electricity generated from renewable sources is also used to process carbon dioxide and water into syngas through the high-temperature electrolysis. This is an attempt to maintain carbon neutral in the generation process. Audi, in partnership with company named Sunfire, opened a pilot plant in November 2014 to generate e-diesel using this process.

Uses

Coal gasification processes to create syngas were used for many years to manufacture illuminating gas (coal gas) for gas lighting, cooking and to some extent, heating, before electric lighting and the natural gas infrastructure became widely available. The syngas produced in waste-to-energy gasification facilities can be used to generate electricity.

Post-treatment

Syngas can be used in the Fischer–Tropsch process to produce diesel, or converted into e.g. methane, methanol, and dimethyl ether in catalytic processes.

If the syngas is post-treated by cryogenic processing, it should be taken into account that this technology has great difficulty in recovering pure carbon monoxide if relatively large volumes of nitrogen are present due to carbon monoxide and nitrogen having very similar boiling points which are −191.5 °C and −195.79 °C respectively. Certain process technology selectively removes carbon monoxide by complexation/decomplexation of carbon monoxide with cuprous aluminum chloride ($CuAlCl_4$) dissolved in an organic liquid such as toluene. The purified carbon monoxide can have a purity greater than 99%, which makes it a good feedstock for the chemical industry. The reject gas from the system can contain carbon dioxide, nitrogen, methane, ethane, and hydrogen. The reject gas can be further processed on a pressure swing adsorption system to remove hydrogen, and the hydrogen and carbon monoxide can be recombined in the proper ratio for catalytic methanol production, Fischer-Tropsch diesel, etc. Cryogenic purification, being very energy-intensive, is not well suited to simply making fuel, because of the greatly reduced net energy gain.

Energy Capacity

Syngas that is not methanized typically has a lower heating value of 120 BTU/scf . Untreated syngas can be run in hybrid turbines that allow for greater efficiency because of their lower operating temperatures, and extended part lifetime.

Synthesis Gas and its Derivatives

Methane and synthesis gas are important petrochemical feedstock for manufacture of a large number of chemicals, which are used directly or as intermediates, many of these products are number of which are finding use in plastic, synthetic fiber, rubber, pharmaceutical and other industries. 'Synthesis gas' is commonly used to describe two basic gas mixtures - synthesis gas containing CO, hydrogen and synthesis gas containing hydrogen and nitrogen for the production of ammonia. Major requirements of synthesis gas in world scale petrochemical are given in Table.

Some of the emerging technologies in utilization of synthesis gas and methane for the production of petrochemicals, are Fischer-Tropsch synthesis, oxidative coupling of methane with chlorine to yield ethane and ethylene, methanol to olefin technology (MTO). Fischer-Tropsch synthesis is being studied in great detail world over and it is promising to be a future technology for manufacture of olefins from synthesis gas. CO that can be separated from synthesis gas either by cryogenic or by pressure swing adsorption is a promising feedstock for production of a variety of products. Product profile of methane, synthesis gas and CO based building blocks are given in Figure.

Table: Synthesis Gas Requirements for Major World Scale Petrochemicals

Product	Required H_2 : CO	Typical world-scale capacity, TPA	Syn. gas required, Nm^3/hr.
Methanol	2:1	1,60,000-12,75,000	48,000-1,90,000
Acetic acid	0:1	2,75,000-5,45,000	18,000-36,000
Acetic anhydride	0:1	90,000	3500
Oxo alcohol	2:1	1,15,000-2,75,000	12,000-25,000
Phosgene	0:1	4,800-1,60,000	3,500-12,000
Formic acid	0:1	45,000	3,500
Methyl formate	0:1	9,000	600
Propionic acid	0:1	45,000-68,000	2,400-3,500
Methyl methacrylate	1:1	45,000	4,700
1,4-Butandiol	2:1	45,000	4,700

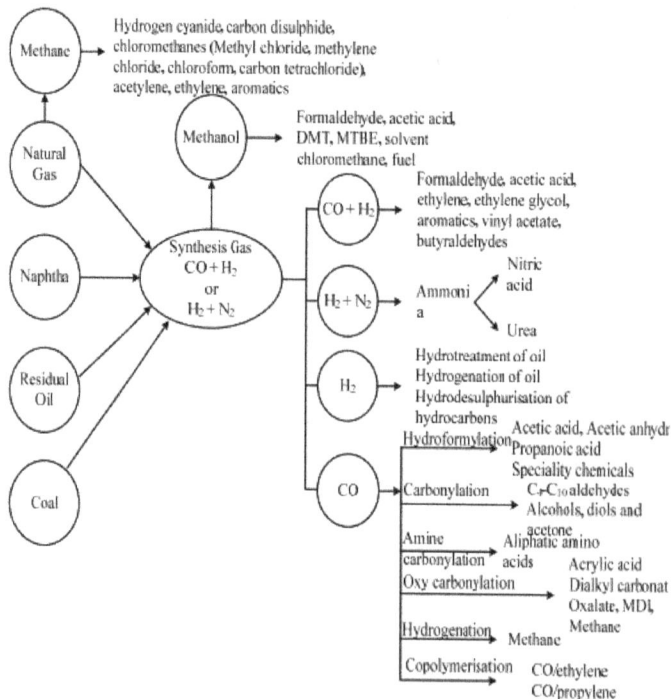

Methane, Synthesis Gas and CO Building Blocks Source: Mall, 2007

Synthesis Gas

Methane and synthesis gas are important petrochemical feedstock for the manufacture of a large number of chemicals, which are used directly or as intermediates, a number of which are finding use in plastic, synthetic fiber, rubber, pharmaceutical and other industries. 'Synthesis gas' is commonly used to describe two basic gas mixtures - synthesis gas containing CO, hydrogen and synthesis gas containing hydrogen and nitrogen for the production of ammonia.

Petrochemical derivatives based on synthesis gas and carbon monoxide have experienced steady growth due to large scale utilization of methanol and development of a carbonylation process for acetic acid and Oxo synthesis process for detergents, plasticizers, and alcohols. Recent market studies show that there will be a dramatic increase in demand of CO and syngas derivatives .

Methanol is the largest consumer of synthesis gas. The reformed gas is to meet certain requirements with regard to its composition. It is characterized by the stoichiometric conversion factor, which differs from case to case

Raw Materials For Synthesis Gas

Various raw materials for synthesis gas production are natural gas, refinery gases, naphtha, fuel oil/residual heavy hydrocarbons and coal. Although coal was earlier used for production of synthesis gas, it has now been replaced by petroleum fractions and natural gas. Petrocoke is the emerging source for Synthesis gas. Coal is again getting importance alone are with combination of petroleum coke. Various Routes for Synthesis gas and Ammonia and Methanol manufacture is shown in Figure. Reactions in the manufacture of synthesis gas by Steam reforming and Partial oxidation in Table.

Process Technology

Various synthesis gas production technologies are steam methane reforming, naphtha reforming, auto-thermal reforming, oxygen secondary reforming, and partial oxidation of heavy hydrocarbons, petroleum coke and coal.

Various steps involved in synthesis gas production through steam reforming are:

- Desulphurization of gas
- Steam reforming and compression
- Separation of CO_2

Various available synthesis gas generation schemes are:

- Conventional steam reforming

- Partial oxidation

- Combined reforming

- Parallel reforming

- Gas heated reforming

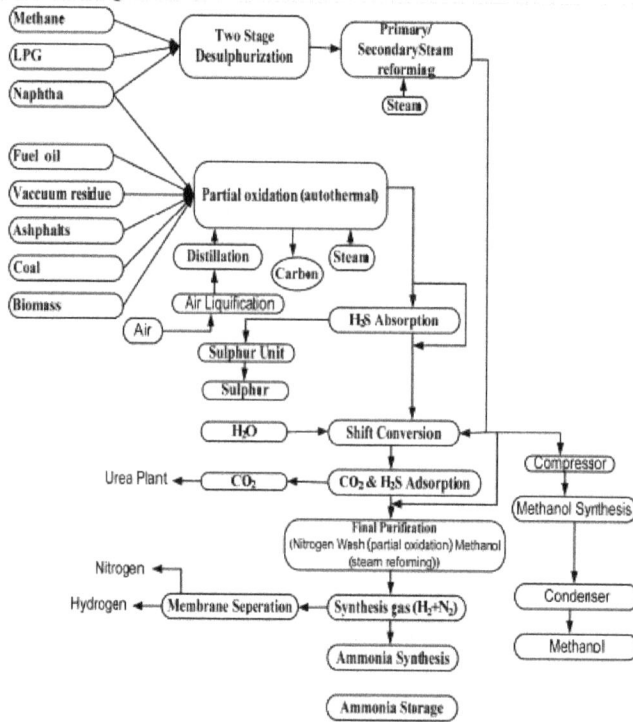

Various Routes for Synthesis gas and Ammonia and Methanol manufacture

Table : Reactions in the manufacture of synthesis gas by Steam reforming and Partial oxidation

Process steps	Reaction	Process Condition
Desulphurisation: 1st Stage First Stage 2nd Stage Second Stage	$C_2H_5SH + H_2 \rightarrow H_2S + C_2H_6$ $C_6H_5SH + H_2 \rightarrow H_2S + C_6H_6$ $C_4H_4SH + 3H_2 \rightarrow H_2S + C_4H_9$ $CS_2 + 4H_2 \rightarrow 2H_2S + CH_4$ $COS + H_2 \rightarrow H_2S + CO$ $CH_3SC_2H_5 + H_2 \rightarrow H_2S + CH_4 + C_2H_4$ $H_2S + ZnO \rightarrow ZnS + H_2O$	Al-Co-Mo Al-Ni-Mo Catalyst 350-400 ^{O}C Zinc oxide absorbent 200-500 ^{O}C

Steam reforming two stages	$C_nH_m + 1/4(4n-m)H_2O \rightarrow 1/8(4n+m)CH_4 +$ $1/8(4n-m)CO_2$ $CH_4 + H_2O \rightarrow CO + 3H_2$ $CO + H_2O \rightarrow CO_2 + H_2$	Nickel catalyst 800 ^oC Endothermic reaction
Partial Oxidation	$C_nH_m + [(2n+m)/4]O_2 \rightleftharpoons nCO + m/2\,H_2O$ $C_nH_m + nH_2O \rightleftharpoons nCO + (n+m/2)\,H_2$ $2CO \rightleftharpoons C + CO_2$ $CO + H_2 \rightleftharpoons C + H_2O$	Exothermic reaction

Methanol

Methanol was first obtained by Robert Boylein in the year 1661 through rectification of crude wood vinegar over milk of lime and was named adiaphorous spiritusliglorum. The term methyl was introduced in chemistry in 1835. Methanol is one of the largest volume chemicals produced in the world. Methanol consumption can be separated into three end use categories – chemical feedstock, methyl fuels, and miscellaneous uses. About 71% of the current global consumption of methanol is in the production of formaldehyde, acetic acid, methyl methacrylate, and dimethyl terephthalate. The global methanol industry has experienced very fundamental and structural changes and has settled down considerably.

Demand changes in key methanol derivatives may adversely affect future demand in case of methanol. Product profile of methanol is given in Table. Globally the demand is expected to grow exponentially, not only caused by growing internal market of traditional applications but accelerated by new applications such as directing blending with gasoline, methanol to olefins (e.g. propylene) and dimethyl ether [Chemical Weekly, November 15, 2011]. Global demand for methanol will reach 122.6 million tones by 2020. Global methanol demand was 26.6 million tons and 44.9 million tones in 2000 and 2010 respectively [Chemical Industry Digest July 2012, p.29]. Present capacity of methanol in India is 4.65 lakh tones. Capacity for methanol and trends in production of methanol is given in Table.

Table : Product Profile of Methanol

Product	Uses
DMT/ Polyethylene terephthalate	Polyester fiber and film, Adhesives, Wire coating, Textile sizing, Herbicides

Methyl methacrylate (MMA)	Cast sheet, surface coating, molding resins, oil additives
MTBE	Oxygenate
Mono methanolamine	Naphthyl-n-methyl carbamate, monoethyl hydrazine, Monomethylamine nitrate
Dimethylamine	Dimethyl acetamide, Dimethyl formamide, Dimethyl hydrazine, 2,4- Di-chlorophenoxyacetic salt
Methylacetate	Paint remover
Dimethylaniline	Solvent, Flavoring, Dyes, Fragrance
Aceticacid	Vinyl acetate, Acetic anhydride, Chloro acetic acid, Ethyl acetate, Butyl acetate, Isopropyl acetate, Acetyl chloride, Acetanilide
Formaldehyde	Phenolic resins, Pentaerythritol, Trioxane, 1,4-butanediol, Formaldehyde, sulphoxylate, Tetraoxane, Resorcinol resin
Methylhalides	Quaternary amines, Methyl cellulose, Butyl rubber, Tri-methanol propene

Table: Profile of Methanol production and Consumption Pattern in India Capacity for Methanol in India

Units	Location	Capacity (Tpa)	Share (%)
Gujarat Narmada Valley Fertilisers Ltd.	Gujarat	238100	51.11
Deepak Fertilisers& Petrochemicals Ltd.	Maharashtra	100000	21.46
Rashtriya Chemicals & Fertilisers Ltd.	Maharashtra	72600	15.58
Assam Petrochemicals Ltd.	Assam	33000	7.11
National Fertilisers Ltd.	Punjab	22110	4.74
Total		465810	100.00

Table: Unit-wise Production and Sales of Methanol

Units	Production		Sales	
	2009-10	2010-11	2009-10	2010-11
Gujarat Narmada Valley Fertilisers Ltd.	187079	202544	111511	126059
Deepak Fertilisers& Petrochemicals Ltd.	65647	81888	65703	81708
Rashtriya Chemicals &Fertilisers Ltd.	44103	68700	19746	41264
Assam Petrochemicals Ltd.	33759	30000	15040	15000
National Fertilisers Ltd.	2669	516	131	44
Total	333257	383648	212131	264075

Table: Methanol Consumption Pattern and Growth

Users	Share (%)	Growth Rate (%)
Formaldehyde	48	7
Pharmaceuticals	21	8.5

Oxygenates	9	-
Acetic Acid	5	4
Alkyl Amines	4	9
Dimethyl Sulphate	3	8
Agrochemicals	3	5
Chloromethanes	4	8
Solvents/Others	3	8
Total	100	6

Methanol Process Technology

From the early 1800s until 1920s, the distillation of wood to make wood alcohol was the source of Methanol. The most common industrially favored method for the production of methanol was first developed by BASF in 1923 in Germany from synthesis gas utilising high pressure process using zinc-chromic oxide catalyst. However, due to high capital and compression energy costs compounded by poor catalyst activity, high-pressure process was rendered obsolete when ICI in the year 1966 introduced a low-pressure version of the process at 5-10 MPa and 210-270 ^{O}C, with a new copper-zinc oxide based catalyst of high selectivity and stability.

Process steps involved in the production of methanol are:

- Production of synthesis gas using steam reforming or partial oxidation

- Synthesis of methanol

- High-pressure process (25 – 30 MPa)

- Medium pressure (10-25 MPa) process

- Low-pressure process (5-10 MPa)

Figure illustrate the production of methanol from steam reforming of natural gas and naphtha.

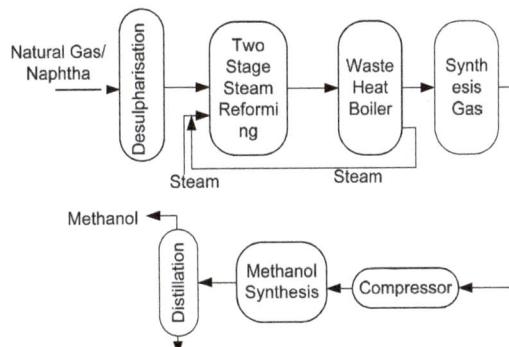

Methanol from steam reforming of Natural gas and Naphtha

The major reactions take place during methanol synthesis converter can be described by following equilibrium reactions:

$$CO + 2\,H_2 \rightarrow CH_3OH \quad \Delta H\,298\,^0K = -90.8\,kJ\,/\,mol$$

$$CO_2 + 3\,H_2 \rightarrow CH_3OH + H_2O \quad \Delta H\,298\,^0K = -49.5\,kJ\,/\,mol$$

$$CO_2 + H_2 \rightarrow CO + H_2O \quad \Delta H\,300\,^0K = 41.3\,kJ\,/\,mol$$

The first two reactions are exothermic and proceed with reduction in volume. In order to achieve a maximum yield of methanol and a maximum conversion of synthesis gas, the process must be effected at low temperature and high pressure.

After cooling to ambient temperature, the synthesis gas is compressed to 5.0-10.0 MPa and is added to the synthesis loop which comprises of following items – circulator, converters, heat exchanger, heat recovery exchanger, cooler, and separator. The catalyst used in methanol synthesis must be very selective towards the methanol reaction, i.e. give a reaction rate for methanol production which is faster than that of competing

Formaldehyde

Some major intermediates derived from formaldehyde are chelating agents, acetal resins, 1,4- butanediol, polyols, methylene diisocynate. It is also used for the manufacture of wide variety of chemicals, including sealant, herbicides, fertilisers, coating, and pharmaceutical. Product profile of formaldehyde is given in Table.

Formaldehyde is commercially available as aqueous solution with concentration ranging from 30-56 wt.% HCHO. It is also sold in solid form as paraformaldehyde or trioxane. The production of formaldehyde in India has been growing at a fairly constant rate during last ten years. There are presently about 17 units in India. Installed capacity and production of formaldehyde during 2003-04 was 2.72 lakh tonnes and 1.89 lakh tonnes respectively.

Various industrial processes for manufacture of formaldehyde using silver and iron- molybdenum catalyst are given below:

Catalyst	Process licensor
Silver catalyst processes	Bayer, Chemical construction, Ciba, DuPont, IG Farben, CdF Chemie process, BASF process, ICI process,
Iron-molybdenum catalyst processes	Degussa process, Formox process, Fischer-Adler, Hiag-Lurgi, IFP-CdF Chimle Lumus, Motedisous, Nikka Topsoe, Prolex

Process diagram for manufacture of formaldehyde using silver and iron-molybdenum catalyst is shown in Figures respectively.

Table: Product Profile of Formaldehyde

Product	Uses
Formaldehyde	Thermosetting resin: Phenol, Urea Melamine, Formaldehyde resins Hexamethy-lenetetramine, Plastic & pharmaceuticals
	1,4-Butadiol
	Methylene diisocyanate
	Fertiliser, Disinfectant, Biocide Preservative, Reducing agent, Corrosion inhibitor
	Polyaceta resin
	p-formaldehyde
	Pentaerythritol (Explosive-PETN),Alkyl resins

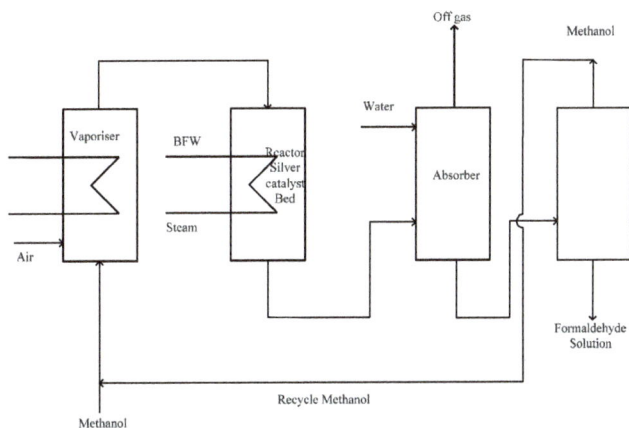

Formaldehyde Using Silver Catalyst

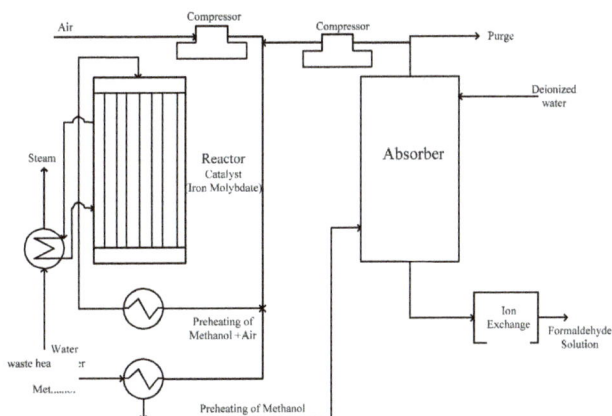

Formaldehyde from Iron Molybdenum Catalyst

Acetic Acid (CH$_3$COOH)

Acetic acid is one of the most widely used organic acid and finds application in the man-ufacture of wide range of chemicals. Acetic acid is the largest methanol based chemical in terms of volume. World capacity and consumption pattern of acetic acid is given in Table. Installed capacity of acetic acid in India is mention in Table. Table shows the market share of acetic acid in India by different companies. Product profile of acetic acid is given in Table.

Table : World Acetic Acid Capacity and Consumption Pattern

Product	Consu mption 2002 ('000 tonnes)	Consumption Growth (percent)			Capacity 2002 ('000 tonnes)	Announced Capacity due by 2012 ('000 tonnes)	Capacity Change Needed by 2012 2002 after An-nouncement	
		1997-2002	2002-2007	2007-2012			('000 tonnes)	percent
Acetic Acid	8,302	3.9	3.4	2.5	9,559	994	1,785	19

Table : Installed Capacities of Acetic Acid in India

Company	Installed capacity (TPA)
Indian Organics Chemicals Ltd.	15,000
Somaiya Chemicals ltd.	15,000
Somaiya organics	20,000
Andhra Sugars Ltd.	1,000
Ashok Organic Industries	30,000
EID Parry (I) Ltd.	10,000
Gujarat Narmada Valley Fertiliser Corp. Ltd.	50,000
Kanoria Chemicals & Industries	6,000
Laxmi Organic Ltd.	9,500
Trichy Distilleries	12,000
Vam Organics	1,15,500
Ashok Alcochem Ltd.	5,400
Dhampur Sugar mills	7,300
Pentokey Ltd.	7,000
Polychem Ltd.	7,500
Trident alcochem	6,000

Table : Market Share of Major Acetic Acid Manufacturer

Name of the companies	Percent Share
Jubilant Organosys Ltd.	22
Ashok Organics ltd.	17
IOCL	9
Gujarat Narmada Valley Fertiliser Corp. Ltd.	9
Others	43

Table: Product Profile of Acetic Acid

Product	Uses
Mono chloro acetic acid	CMC manufacture, adhesives, thickeners for drilling muds, food industry, pharmaceuticals, textiles, 2,4-D(insecticides)
Ethyl acetate,n-butyl acetate, isopropyl acetate	Coatings, adhesives, inks and cosmetics
Vinegar	Food Preservative
Cellulose Acetate	Fibers, plastic film
Acetic anhydride	Pharmaceuticals, intermediates, cellulose acetate
Acetanilide	Pharmaceutical, dyes intermediate, Rubber accelerator, Peroxide stabilizers
Per acetic acid	Special Oxidants
Terephthalic Acid, DMT	Polyester fiber, packaging, photographic films, magnetic tape sectors
Vinyl acetate	Polyvinyl acetate, polyvinyl chloride, paints, Adhesives, and coatings

Chloromethanes (Methyl Chloride, Methylene Dichloride, Chloroform, Carbon Tetrachloride)

Chlorinated methanes, which include methyl chloride, methylene dichloride, chloroform and carbon tetrachloride, are important derivatives of methane and find wide application as solvents and as intermediate products. Product profile of Chloromethane is given in Table.

Table: Product Profile of Chloromethane

Product	Uses
Methyl chloride	Refrigerant, butyl rubber, silicones, solvent, tetramethyl lead, intermediates
Methylene dichloride	Solvent, Intermediates, Photographic film, Degreasing solvents, Aerosol, Propellants
Chloroform	Chlorodifluoro methane, (Refrigerants), Propellants, Pharmaceuticals
Carbon tetra chloride	Dichlorodifluoro methane, Trichlorofluoro methane, Solvent, Fire extinguishers

Process Technology

There are two major routes for the manufacture of chloromethane:

- Direct chlorination of methane

- Through methanol route

Direct Chlorination of Methane: Chlorination of methane (natural gas) is carried out at around 400-450 ^{O}C during which following reaction takes place:

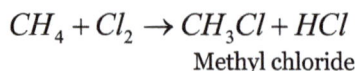

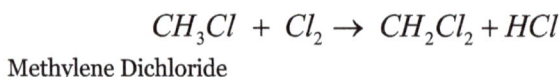

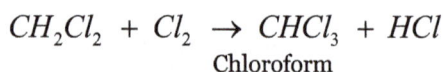

$$CH_4 + Cl_2 \rightarrow CH_3Cl + HCl$$
Methyl chloride

$$CH_3Cl + Cl_2 \rightarrow CH_2Cl_2 + HCl$$
Methylene Dichloride

$$CH_2Cl_2 + Cl_2 \rightarrow CHCl_3 + HCl$$
Chloroform

$$CHCl_3 + Cl_2 \rightarrow CCl_4 + HCl$$

Dimethyl Formamide [HCON(CH$_3$)$_2$]

Dimethyl formamide is one of widely used solvents in the manufacture of acrylic fiber. Because of its high dielectric constant, aprotic nature, wide liquid range and low volatility, dissolving power it is frequently used for as solvent.

Process Technology

Dimethyl formamide is made by following two processes:

Two Step Process

Process involves carbonisation of methanol to methyl formate using basic catalyst and reaction of methyl formate with dimethylamine.

$$CH_3OH + CO + HCOOCH_3 \rightarrow (CH_3)_2 NH + HCON(CH_3)_2 CH_3OH$$

Acetylene

It plays important role during and after World War II in providing feedstock for large number of organic chemicals when petrochemical industry was not well developed. Acetylene's highly reactive triple bond provided a ready "handle" for chemists to grab onto for designing process chemistry

Safety issues involved with handling of large volumes of acetylene and its expense are big problem with adoption of acetylene based processes. The process of acetylene requires much energy and is very expensive. of attractive petrochemical feedstock. Acetylene is still being used for manufacture of chemicals.

Various Routes for Acetylene

Calcium Carbide Route: This is the oldest method for production of acetylene and still acetylene is produced by this process in small scale as well large scale. Calcium carbide is produced by reacting lime with coke at temperature 2,000-2,100°C in an electric furnace. Two processes produce acetylene from calcium carbide process: Wet process and Dry process. Dry process is preferred as in case of calcium hydroxide, which is produced during the process (is produced in the form of dry calcium hydrate).

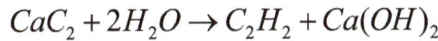

$$CaC_2 + 2H_2O \rightarrow C_2H_2 + Ca(OH)_2$$

Acetylene from Cracking of Hydrocarbons: Cracking of hydrocarbons such as methane, ethane, propane, butane, ethylene, and natural gas can make acetylene.

- $2CH_4 \rightarrow C_2H_2 + 3H_2$

- $C_2H_4 \rightarrow C_2H_2 + H_2$

- $C_4H_{10} \rightarrow C_2H_2 + C_2H_4 + 2H_2$

Product Derived from Acetylene: Acetylene is extremely reactive hydrocarbon and was initially was used for the manufacture of large number of chemicals which are now being derived from acetylene route. Product profile of acetylene is given in Figures.

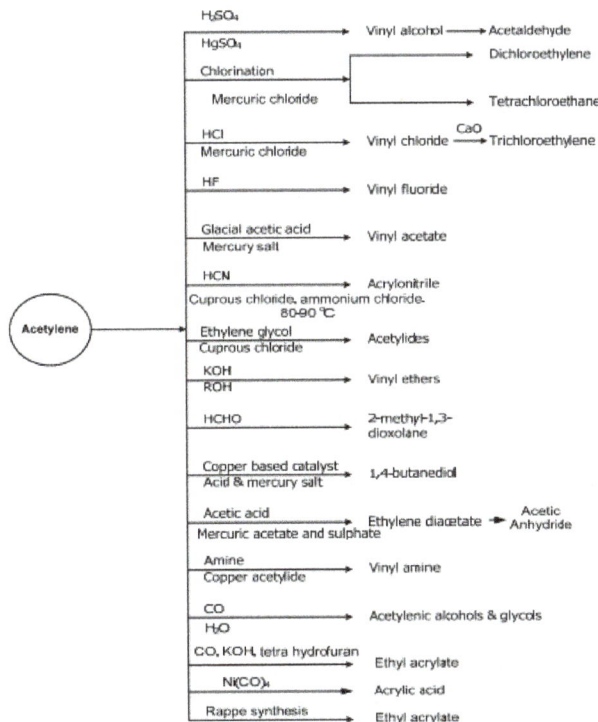

Product Profile of Acetylene

Acetaldehyde:

$$HC \equiv CH + H_2O \rightarrow CH_3CHO$$

Acrylonitrile:

$$HC \equiv CH + HCN \rightarrow HC = CH = CH_2 = CHCN$$

Chlorinated solvents:

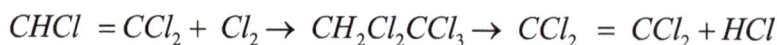

$$HC \equiv CH + 2Cl_2 \rightarrow CH_2Cl_2CHCl_2 \rightarrow CHCl = CCl_2 + HCl$$

$$CHCl = CCl_2 + Cl_2 \rightarrow CH_2Cl_2CCl_3 \rightarrow CCl_2 = CCl_2 + HCl$$

Vinyl acetate:

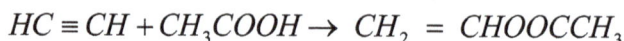

$$HC \equiv CH + CH_3COOH \rightarrow CH_2 = CHOOCCH_3$$

Chloreprene

$$HC \equiv CH + CH_3COOH \rightarrow CH_2 = CHOOCCH_3$$

$$CH_2 = CHOOCCH_3 + Cl_2 \rightarrow CH_2 = CClCH = CH_2$$

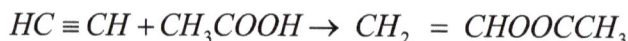

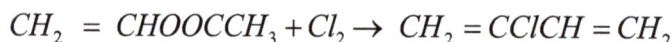

Vinyl Chloride and Vinylidene Chloride

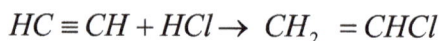

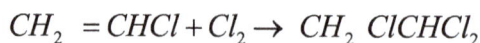

$$HC \equiv CH + HCl \rightarrow CH_2 = CHCl$$

$$CH_2 = CHCl + Cl_2 \rightarrow CH_2 ClCHCl_2$$

$$CH_2 ClCHCl_2 \rightarrow CH_2 = CCl_2 + HCl$$

Vinyl fluoride:

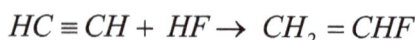

$$HC \equiv CH + HF \rightarrow CH_2 = CHF$$

Reactions in Acetylene derived Chemicals

Ethylene Derivatives

Ethylene is one of the most versatile petrochemicals and its production has steadily increased over the years. Ethylene is called as king of chemicals and surpasses all organic chemicals in production and in amount sold. Ethylene is the basic building block for petrochemicals. Product profile of ethylene is given in Table. Because of its ready

availability at low cost and high purity & reactivity, ethylene has become one of the important raw materials for large number of petrochemicals and products. Ethylene has replaced the earliest route of production of vinyl chloride, acetaldehyde, vinyl acetate and other chemicals through acetylene route. Installed and production capacity of ethylene and its derivatives is mention in Table. Large tonnage of ethylene is being used for the manufacture of polyethylene, ethylene oxide, ethylene glycol and styrene. World requirement of ethylene is given in Figure. World ethylene capacity is shown in Figure. World ethylene complexes, capacity and top 10 producers are mention in Table respectively. Global ethylene capacity growth is given in Table.

Table: Product Profile of Ethylene

Product	Uses
Polyethylene LDPE, LLDPE, HDPE	Films, moldings, pipes, cable covering, netting
Ethylene oxide & Ethylene glycol	Antifreeze, polyester, solvents, detergent, textile, ethanol amine
Styrene	Synthetic rubber and polystyrene
Ethyl alcohol	Industrial solvent and Chemical intermediates
Acetaldehyde (from ethyl alcohol)	Acetic acid (Peracetic acid), Acetic anhydride, Cellulose acetate, Vinyl acetate, Pyridine, Butyraldehyde (ethyl hexanol)
Olefin	n-Butenes, Synthetic detergent, Oxo alcohols, Synthetic lubricants
Chlorinated Solvents	Trichloroethylene, Perchloroethylene
Ethyl Chloride	Tetraethyl lead, Chemical intermediates
Vinyl Acetate	Polyvinyl acetate, Polyvinyl alcohol
Vinyl Chloride	Polyvinyl chloride(PVC)

Table: Installed Capcity and production of Important Ethylene & Ethylene Derivatives in India

Product	Installed capacity 2008-09, "000"MT	Production 2009-10, "000" MT
Ethylene	2841	2515
Ethylene oxide	120	117
Monoethyl;ene glycol	820	738
Acetaldehyde	238	59.2
Acetic anhydride	59	43.42
Ethanol amine	10	7.0
Ethyl acetate	132.0	103.96

Table: World Top Ethylene Complexes

	Company	Location	Capacity, TPY
1	Formosa Petrochemical Corp.	Mailiao, Taiwan, China	2935000
2	Nova Chemicals Corp.	Joffre, Alta	2811792
3	Arabian Petrochemical Co.	Jubai, Saudi Arabia	2250000
4	Exxon Mobi Chemical Co.	Baytown, Tex.	2197000
5	Chevron Philips Chemical Co.	Sweeny, Tex.	1865000
6	Dow Chemical Co.	Terneuzen, Netherlands	1800000
7	Ineos Olifins& Polymers	Chocolate Bayou, Tex.	1752000
8	Equistar Chemicals LP	Channelview, Tex.	1750000
9	Yanbu Petrochemical Co.	Yanbu, Saudi Arabia	1705000
10	Equate Petrochemical Co.	Shuaiba, Kuwait	1650000

Table: Regional Capacity Breakdown

	Ethylene Capacity, tpy
Asia-Pacific	42631000
Eastern Europe	7971000
Middle East, Africa	23357000
North America	34508000
South America	5083500
Western Europe	24904000
Total Capacity	138454500

Table: Top 10 Ethylene Producers

Company	Sites	Capacity, TPY	
		Of Entire Complexes	With Only Company Partial Interests
Saudi Basic Industries Corp.	15	13392245	10273759
Dow Chemical Co.	21	13044841	10529421
ExxonMobil Chemical Co.	19	12515000	8550550
Royal Dutch Shell PLC	13	9358385	5946693
Sinopec	13	7575000	7275000
Total AS	11	5933000	3471750
Chevron Philips Chemicals Co.	8	5607000	5352000
Lyondell Basell	8	5200000	5200000
National Petrochemical Co.	7	4734000	4734000
Ineos	6	4656000	4286000

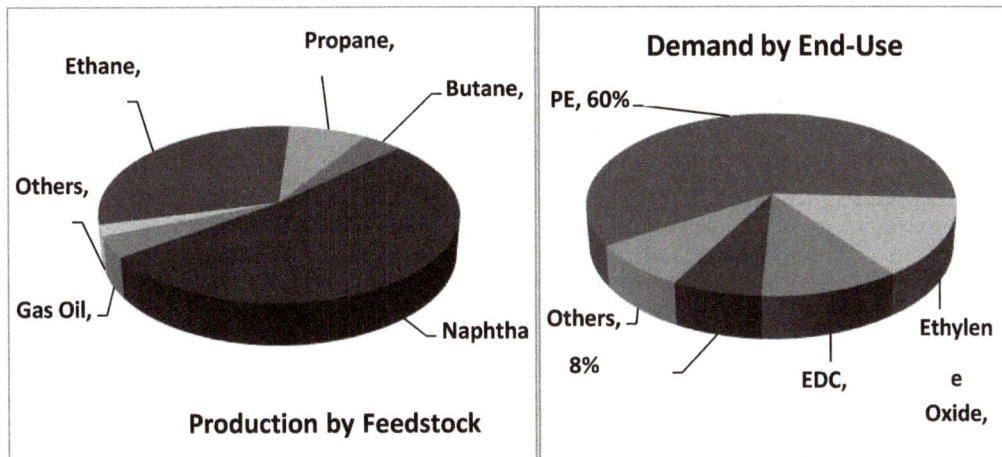

World Ethylene Supply/Demand Profile 2008 production = 117.3 million metric tons

Table: Global Ethylene Capacity Growth (-000-Tons)

Major Region	2008 Capacity	2013 Capacity	08 to '13 Delta
Middle East / Africa	19,711	34,461	14,751
Asia Pacific	39,617	56,349	16,732
America's	40,421	40,434	12
Europe	30,953	31,293	340
World Total	130,702	162,537	31,836

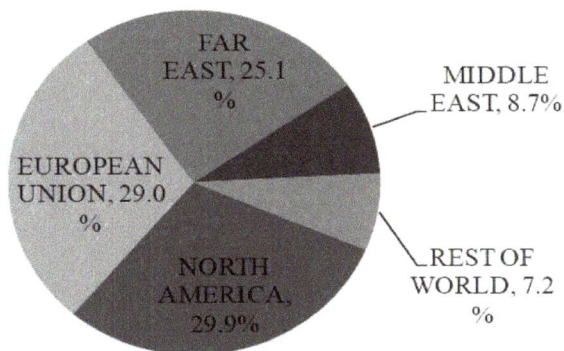

World Ethylene Capacity (120 Million tonnes 2008)

Etylene Oxide (EO)

Ethylene oxide is one of the important petrochemical intermediate used for the manufacture of large number of products; some of the major uses are in the manufacture monoethylene glycol, glycol ethers which is made by reaction of ethylne oxide and alcohols, ethanol amines. Surfactant indutry is one of largest user of EO, both for industrial and house hold applications. Product profile of ethylene oxide is given in Table.

Table: Product Profile of Ethylene Oxide

Product	Uses
Ethanol Amines	Detergent, Soap solvent, Cosmetics, Morpholine
Mono Ethylene Glycol	Polyester, Staple fiber yarn, PET Bottles, Film
Diethylene Glycol	Coolants, Pesticides, Rubber Compounding, Plasticizer, Polyurethane, Alkyl Resin
Tri ethylene Glycol	Natural gas conditioning agents plasticizer
Polyethylene Glycol	Pharmaceuticals, Brake Fluid, cosmetics
Non-ionic surfactants, Ethoxylates	Textile auxiliaries, Binders, Dyes, Pesticides, Pharmaceuticals, Cosmetics
Glycol Ethers	Brake Fluid and Protective Coating

Chlorohydrin Process

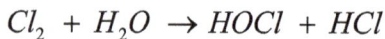

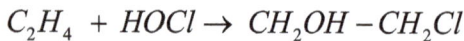

$$Cl_2 + H_2O \rightarrow HOCl + HCl$$

$$C_2H_4 + HOCl \rightarrow CH_2OH - CH_2Cl$$

$$CH_2OH - CH_2Cl + Ca(OH)_2 \rightarrow CH_2 - CH_2 + CaCl_2 + H_2O$$

Direct Oxidation

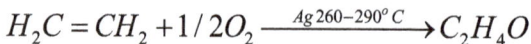

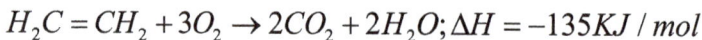

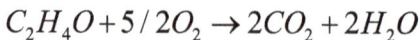

$$H_2C = CH_2 + 1/2O_2 \xrightarrow{Ag\ 260-290^\circ C} C_2H_4O$$

$$H_2C = CH_2 + 3O_2 \rightarrow 2CO_2 + 2H_2O; \Delta H = -135KJ/mol$$

$$C_2H_4O + 5/2O_2 \rightarrow 2CO_2 + 2H_2O$$

$$H_2C = CH_2 + 1/2O_2 \rightarrow CH_3CHO$$

$$H_2C = CH_2 + 2O_2 \rightarrow HCHO$$

MONO-, DI- TRI- ETHYLENE GLYCOLS (MEG, DEG, TEG)

A major petrochemicals and find application in manufacture of polyester and as antifreeze accounts for 70% of Ethylene oxide production. Ethylene oxide preheated to 1950C. EO:H2O ratio 10:1 to maximize MEG production By Products DEG, TEG. Figure gives detail manufacturing of MEG, DEG and TEG from Ethylene Oxide

$$C_2H_4O + H_2O \longrightarrow HO-\underset{H_2}{\overset{H_2}{C}}-C-OH$$
$$MEG$$

$$HO-\underset{H_2}{\overset{H_2}{C}}-C-OH + C_2H_4O \longrightarrow HO(C_2H_4O)_2H$$
$$MEG \qquad\qquad\qquad\qquad DEG$$

$$HO(C_2H_4O)_2H + C_2H_4O \xrightarrow{332} HO(C_2H_4O)_3H$$
$$DEG \qquad\qquad\qquad\qquad TEG$$

MEG, DEG and TEG from Ethylene Oxide

Vinyl Chloride

Vinyl chloride is one of the important petrochemical feedstocks and find use in manufacture of poly vinyl chloride the second largest tonnage commercial polymer after polyethylene. About 95 percent of the present vinyl chloride production worldwide is used in polymer production or copolymer application. Another important use of vinyl chloride is in the production of vinylidiene chloride. According to SRI consulting global production and consumption of Ethylene dichloride (EDC) in 2009 (which accounts for 95 percent consumption in vinyl chloride manufacture), was about 33.7 million tones with global capacity of about 73 percent in 2009.

Process Technology

The original process of manufacture of vinyl chloride was by reaction of acetylene derived from calcium carbide with hydrochloric acid in gaseous phase in presence of mercuric chloride catalyst at temperature around 100-180 °C. However with the availability of ethylene from cracker plant now vinyl chloride is made from ethylene obtain from cracker plant.

Direct Chlorination

- The process of vinyl chloride manufacture takes place in two stages.

- First stage: Ethylene is reacted with chlorine in either liquid or vapor phase in presence of ferric chloride. However, the liquid phase process is more common and the reaction takes place at around 50-90 °C and 3-5 atm pressure.

- Second stage: Vinyl chloride is produced by pyrolysis of vaporised ethylene dichloride in a set of tubular furnaces at temperature of about 400-500 °C.

- Direct chlorination:

$$CH_2 = CH_2 + Cl_2 \rightarrow ClCH_2 - CH_2 Cl$$

Process Technology

The process of vinyl chloride manufacture takes place in two stages.

First Stage: Ethylene is reacted with chlorine in either liquid or vapor phase in presence of ferric chloride. However, the liquid phase process is more common and the reaction takes place at around 50-90 °C and 3-5 atm pressure.

Second Stage: Vinyl chloride is produced by pyrolysis of vaporised ethylene dichloride in a set of tubular furnaces at temperature of about 400-500 °C.

Ethylene chloride by direct Chlorination of Ethylene:

- The original process of manufacture of vinyl chloride by ethylene chlorination and cracking of ethylene dichloride had been replaced by oxychlorination process in which no hydrochloric acid is formed as byproduct. Process diagram of vinyl chloride from oxychlorination process is shown in Figure.

- The process involves production of ethylene dichloride by exothermic reaction of ethylene, hydrochloric acid and oxygen

- Liquid phase: Fixed or fluidized bed reactor is used at 170-180 oC and 15-20 atm pressure in presence of copper chloride.

- Vapor phase reaction: The temperature and pressure are 200-220 oC and 20-50 atm pressure.

Reactions

Vinyl Chloride by Chlorination

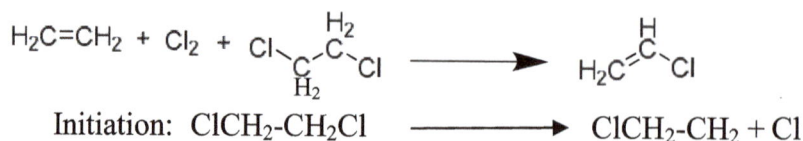

Initiation: $ClCH_2\text{-}CH_2Cl \longrightarrow ClCH_2\text{-}CH_2 + Cl$

Propagation:

$$Cl + ClCH_2 - CH_2Cl \rightarrow ClCH_2 - CHCl + HCl$$

$$ClCH_2 - CHCl \rightarrow CH_2 = CHCl + Cl$$

Termination:

$$Cl + ClCH_2 - CH_2 \rightarrow CH_2 = CHCl + HCl$$

The first stage is typical electrophilic addition of a halogen to an alkene. The second stage is a free radical chain reaction.

Oxychlorination

The original process of manufacture of vinyl chloride by ethylene chlorination and cracking of ethylene dichloride had been replaced by oxychlorination process in which no hydrochloric acid is formed as byproduct. The process involves production of ethylene dichloride by exothermic reaction of ethylene, hydrochloric acid and oxygen. Liquid phase: at about 170-180 °C in at 15- 20 atm pressure in presence of copper chloride in either fixed or fluidised bed reactor. Vapor phase reaction: the temperature and pressure are 200-220 °C and 20-50 atm pressure.

- Direct chlorination: $CH_2 = CH_2 + Cl_2 \rightarrow ClCH_2 - CH_2Cl$

- Oxychlorination: $CH_2 = CH_2 + 2\ HCl + \frac{1}{2}\ O_2 \rightarrow ClCH_2 - CH_2Cl + H_2O$

- Ethylene dichloride pyrolysis: $ClCH_2 - CH_2Cl \rightarrow CH_2 = CHCl + HCl$

- Overall reaction: $2\ CH_2 = CH_2 + Cl_2 + \frac{1}{2}\ O_2\ 2 \rightarrow CH_2 = CHCl + H_2O$

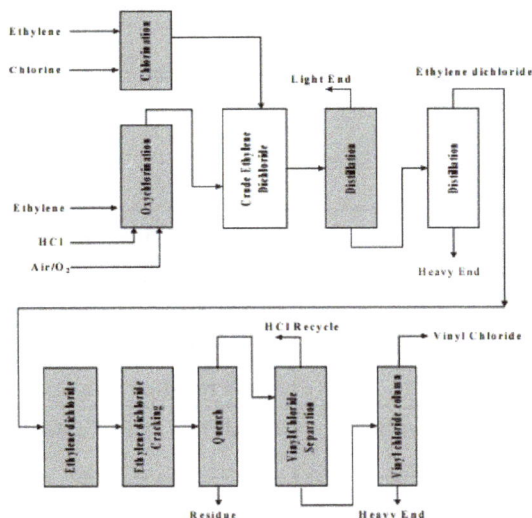

Vinyl Chloride from Oxychlorination Process

Vinyl Acetate

Vinyl acetate is one of the important derivatives of ethylene which is used as intermediate for manufacture of polyvinyl alcohol, polyvinyl acetate, polyvinyl butyral, etc.

Global end use pattern of vinyl acetate is:- Adhesives (23%), paints and coating (29%), textiles (21%), plastics (17%), paper and board (10%). Consumption pattern of vinyl acetate in India is polyvinyl acetate emulsions & resins (50%), polyvinyl alcohol (25%), ethylene vinyl acetate (10%), others (15%).

End use pattern of vinyl acetate in India is :- Adhesives (35-40%), textiles (30-35%), paints and coating (15-20%), others (10-15%)

Process Technology:

- The ethylene route has replaced the traditional process of manufacture of vinyl acetate. The production of vinyl acetate through acetylene route, which was developed by Wacker in 1930, involves reaction of acetylene and acetic acid in liquid phase at 60-80 °C and 1-2 atm pressure in presence of mercury salt catalyst.

$$H_2C = CH_2 + CH_3COOH \rightarrow H_3COOCHC = CH_2$$

Vinyl Acetate from Ethylene Route

- Vinyl acetate is made by reaction of ethylene with acetic acid by liquid phase process or by Vapor phase process in presence of palladium and cupric chloride catalyst. In the vapor Phase process, following reactions take place:

$$CH_2 = CH_2 + CH_3 COOH + PdCl_2 \rightarrow CH_2 = CHCOOCH_3 + 2HCL + Pd$$

$$Pd + 2CuCl_2 \rightarrow PdCl_2 + 2CuCl$$

$$2CuCl + 2HCl + H_2O_2 \rightarrow 2CuCl_2 + 2H_2O$$

Use of Vinyl Acetate

Vinyl Acetate	Polyvinyl Acetate	Surface coating adhesives, Textile resins
	Polyvinyl Alcohol	Textile size, Grease proofing paper, Vinyl emulsifier, Thickener, Viscosity regulators, Adhesives,
	Acrylo- nitrile Copolymer	Acrylic Fibers
	Polyvinyl Formate	Water resistant insulation enamel
	Ethylene Vinyl Acetate Copolymers	Textile and Paper Coating
	Vinyl chloride comonomers	VC-VAC, LP Records, VC-VAC Coating
	Polyvinyl Butryaldehyde	Safety Glass

Ethanol

Ethanol apart from its major use as a beverage is one of the most versatile chemicals and is one of the basic building blocks of the organic chemical industry. Ethanol is generally produces by fermentation of molasses, due to the development of petrochemical industry and availability of ethylene, now ethylene provides another major route of formation of ethanol.

However, still molasses were used to produce ethanol in India. In India some of the important chemical are still prepared through ethanol which were earlier prepared through petrochemical route. Two such important complexes are Jubilant Organosys Ltd., Gajraula (Uttar Pradesh) and Indian Glycol Ltd., Kashipur (Uttar Pradesh), where large number of ethanol derivatives are manufactured through ethanol route.

Various routes for manufacture of ethanol

- Fermentation of molasses
- Catalytic hydration of ethylene
- Ethylene esterification and hydrolysis

Fermentation of Molasses:

Ethyl alcohol is prepared from molasses by fermentation process utilising yeast enzymes. Separation of 8-10% alcohol is achieved in a series of distillation columns, as alcohol and water at 95% concentration form azeotropic mixture.

Ethanol by Esterification and Hydrolysis

- Ethylene and sulphuric acid are reacted at 80 oC and 1.5 MPa to form a mixture of ethyl sulphates, which are then hydrolysed to ethyl alcohol.
- Ethylene and sulphuric acid are reacted in absorber from which the mixture of ethylene sulphates thus formed is fed to hydrolyser from which the crude alcohol and sulphuric acid are fed to stripping section and caustic scrubbing section and finally to a series of two distillation columns for separation of ether and alcohol.

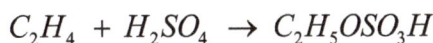

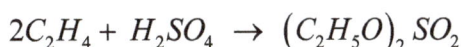

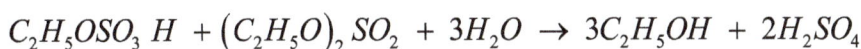

$$C_2H_4 + H_2SO_4 \rightarrow C_2H_5OSO_3H$$

$$2C_2H_4 + H_2SO_4 \rightarrow (C_2H_5O)_2 SO_2$$

$$C_2H_5OSO_3 H + (C_2H_5O)_2 SO_2 + 3H_2O \rightarrow 3C_2H_5OH + 2H_2SO_4$$

$$2C_2H_5OH \rightarrow C_2H_5OC_2H_5 + H_2O$$

Ethanol by Vapor Phase Hydration of Ethylene

- An ethylene rich gas is mixed with water and heated to about 300 oC and passed on to fixed bed catalytic reactor where catalytic hydration of ethylene takes place

$$C_2H_4 + H_2O \rightarrow C_2H_5OH$$

- The catalyst used is phosphoric acid deposited on silica gel. The reactor effluents are sent to separator for separation of vapor and liquid. The gases from the separator are cooled and scrubbed with water to recover traces of alcohol. The alcohol water mixture is sent to a series of distillation columns where ether is separated in the light end column and finally 95% by volume ethanol water azeotrope is separated.

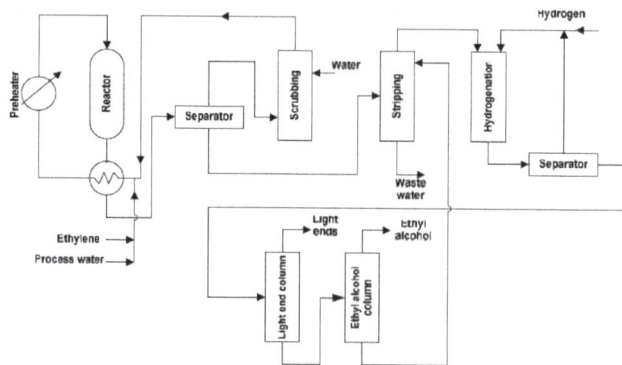

Ethanol from Catalytic Recycle Hydration of Ethylene

Acetaldehyde

Table: Product Profile of Acetaldehyde

Product	Uses
Pyridine, Picoline	Solvent, Drugs, Dyes, Agricultural chemicals
Chlor-aldehydes	Insecticides, Fungicides, Disinfectants
Acetaldol	1,3-Butylene Glycol (Polyesters),Urethane coating, Humcetant, Printing ink,Crotonaldehyde,n-butyl alcohol, n-butyric acid anhydride, 2-ethyl hexanol ,Rubber accelerator, Sorbic acid
Paraldehyde	Rubber accelerator, Antioxidant dye,stuff
Per-acetic acid	Epoxidation reaction, Reagent in caprolactam, Synthetic glycerols
Penta-erythritol	Alkyl resin, Stabilizer, Plasticizers, Chlorinated polyether resin, Intumescents
Acetic, anhydride	Acetyl salicylic acid, Cellulose acetate, Esters
Acetic acid	Cellulose acetate, Vinyl acetate Chloro-acetic acid Ammonium acetate
Lactic acid	Food and, beverages, lactates, adhesives, Leather processing

Propylene

Propylene often referred as the crown prince of petrochemicals is superficially similar to ethylene but there are many differences in both production and uses. Propylene is used in many of the world's largest and fastest growing synthetic materials and thermoplastics. The demand of propylene has increased rapidly during the last twenty years and primarily driven by polypropylene demand. Product profile of propylene is given in Table.

According to SRI consulting 2010 global production and consumption of propylene in 2009 was both approximately 71 million tones with capacity utilization of 78.5%. Global propylene consumption is forecast to average growth of around 5.1% per year from 2009 to 2014 and 3.5% per year from 2014-19. Consumption of refinery grade propylene made up 9% of total consumption in 2009, chemical grade 23% and polymer grade 68%. Refinery grade propylene is consumed mainly for production of cumene and isopropyl alcohol. Chemical grade propylene mostly goes into oxo alcohol, propylene oxide and acrylonitrile.

Table: Product Profile of Propylene

Product	Uses
Miscellaneous chemicals	1 butanol, 2-ethyl hexanol, Allyl chloride, Epichlorohydrin
Polymer	Polypropylene, Polyacrylamide, nylon 66, acrylic sheets
Propylene oxide	Polyether-polyols, glycol ethers, isopropyl amines, propylene carbamate, surfactants
Propylene glycol	Unsaturated polyester resins, food additives, cellophane, paints and coating, plasticisers, functional fluids, antifreeze, tobacco treatment
Acrylonitrile	Acrylic fiber, acrylic acid, acrylates, methyl methacrylates, adiponitrile
Isopropanol	Acetone, cosmetics, solvents, pharmaceuticals, isopropyl acetate
Polyols	Polyurethane and Polyester

Sources of Propylene

Propylene is a byproduct of steam crackers and varying amount of olefins is produced from steam crackers depending on the type of feedstock. Other sources of propylene may be recovery of propylene from FCC light ends, Propane dehydrogenation, Metathesis. Some of the major processes for production of propylene is given in Table. Typical Composition of FCC Gas Stream is given in Table.

Table: Propylene Production Technologies

Technology	Process	Licensor
Olefin conversion technology	This process involves production of propylene from ethylene and 2-butenes in a fixed bed metathesis reactor containing proprietary catalyst, which promotes reaction of ethylene and 2-butene to form propylene and simultaneously isomerises 1-butene to 2-butene.	ABB Lumus Global

Superflex Process	The process uses a fluidised bed catalytic reactor system using proprietary catalyst which converts low value feedstock to predominantly propylene and ethylene products. Low value light hydrocarbon streams from ethylene plant and refineries can be used, e.g. C_4 and C_5 olefin rich stream from ethylene plants, FCC naphtha, C_4 stream, thermally cracked naphtha from visbreakers or cokers.	Kellogg Brown & Root, Inc.
Propylur Process	This process produces propylene beside ethylene from low value rich feeds ranging from C_4-C_8 from ethylene plant and refineries in a fixed bed reactor using proprietary catalyst. The process offers high selectivity towards propylene.	Lurgi Oel Gas Chemie GmbH
UOP Oleflex Process	This process produces polymer grade propylene from propane and the process consist of a reactor, catalyst regeneration section and product separation and fractionation section. The process uses platinum catalyst (DeH-12 catalyst).	UOP LLC
UOP/Hydro MTO Process	This process converts crude methanol (produced from synthesis gas using natural gas) to ethylene and propylene and can be operated either a maximum ethylene or a maximum propylene production mode using MTO-100 silicoaluminophosphate synthetic molecular sieve based catalyst. The process utilizes fluidised bed reactor and regenerator.	UOP LLC and Hydro Norway
Methanol to propylene (MTP) Technology	This process produces propylene through methanol route using natural gas. In this process propylene is produced in two steps. First methanol is converted to dimethyl ether in reactor followed by reaction of methanol/DME in second reactor. Methanol can be produced from methane from conventional method.	Lurgi Oel Gas Chemie GmbH
C4 hydrogenation and Meta-4 Process	This process involves production of polymer grade propylene plus an isobutylene rich stream or MTBE by upgrading low value C_4 stream pyrolysis C_4 cuts or butene rich cut. The process steps involve – butadiene and C_4 acetylenes selective hydrogenation and butadiene hydroisomerisation, isobutylene removal or MTBE production and metathesis step for conversion of butene and ethylene to propylene. The two main equilibrium reactions taking place are metathesis and isomerisation.	Axens, Axens NA
Olefin Ultra™	A new ultra high activity ZSM-5 additive that provides the highest activity has been developed by Davision catalysts.	
KBR's MAXOFIN-3 TECHNOLOGY	KBR's MAXOFIN process is based on fluidised bed cracking of gas oils and residue feeds using ZSM catalyst and proprietary MAXOFIN-3 catalyst additive. The process gives 15% or higher propylene yield from gas oil.	Kellogg Brown & Root, Inc.

Table: Typical Composition of FCC Gas Stream

Products	Yield weight (percent)
Dry gas (including ethylene)	12.7
Propane	6.5
Propylene	21.0
Butene	35.8

Catalytic Dehydrogenation

Catalytic dehydrogenation of light paraffins is of increasing importance because of the growing demand of olefins such as propylene and isobutene [Reasco and Haller, 1994] and n- butenes. Propane dehydrogenation accounts for 2percent of the total world propylene production. Some of the commercial processes available for dehydrogenation of propane and n-butane are [Badoni et al., 1996]:

- Oleflex (UOP).

- Catofin (ABB Lumus).

- FBD-4 (Snamprogetti SPA).

- Star (Phillips Petroleum Company).

Catalytic dehydrogenation takes place at high temperature (650 °C) using platinum based or chromium-alumina or Fe, Cr/Al_2O_3 as catalyst. Reactor effluent treatment for the separation of hydrogen, propylene, and propane is not simple and total investment is high. These production units can be installed only in areas where field propane is available at low costs.

Methanol to Propylene:

This process produces propylene from natural gas via methanol by converting methanol to dimethyl ether in adiabatic reactor using high activity, high selectivity catalyst. The methanol, water, DME stream is then feed to series of MTP reactor where steam is added. The product stream is first processed for removal of traces of water, CO_2 and DME, followed by further processing for yielding polymer grade propylene.

Propylene Oxide, Propylene Glycol and Polyols

Propylene oxide, propylene glycols and polyols are important derivatives of propylene. propylene oxide is used for the manufacture of propylene glycol and polyols. Major consumption of propylene oxide is manufacture of polyurethane and polyester resins. Propylene glycol find major application in the manufacture of unsaturated polyester resins, food additives, pharmaceuticals and personal care, tobacco humectants, cellophane, paints and coatings. Polyols major use is in the manufacture of polyurethane.

Propylene Oxide

Various rote for making propylene oxide are

There are two major processes for the manufacture of propylene oxide: Propylene chlorohydrin process and propylene oxidation process using peroxides.

Propylene Chlorohydrin Route:

The chlorohydrination process consists of formation of propylene chlorohydrin by the reaction between hypochlorous acid and propylene. The propylene chlorohydrin is epoxidised to propylene oxide by a 10% solution of milk of lime or NaOH. Various steps involved are

- Propylene hypochlorination: Propylene is reaxted with aquous chlorine resulting in the formation of propylene chlorohydrins. Unreacted propylene is recyled.

- Neutralisation: Neutralisation of propylene chlorohydrins containing hydrochloric acid which is formed during the process.

- Dehydrochlorination: Reaction of propylene chlorohydrin with milk of lime or caustic soda to produce propylene oxide

- Purification: Distillation of crude propylene oxide for separation heavy ends

Reactions:

$$Cl_2 + H_2O \longrightarrow HOCL + HCL$$

$$CH_3 - CH = CH_2 + HOCl \xrightarrow{350\,^\circ C} CH_3CHOH - CH_2 \Delta H^\circ_{298} = -225KJ\,/\,mol$$

Propylenechlorohydrin

$$2CH_3CHOH - CH_2Cl + Ca(OH)_2 \text{ or } NaOH \longrightarrow 2CH_3\,CH - CH_2 + CaCl_2 \text{ or } NaCl + 2H_2O\,\Delta H^\circ_{298} = 5KJ\,/\,mol$$

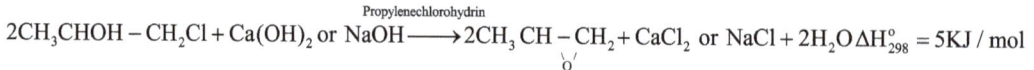

Byproducts formed during the reaction are 1,2-dichloropropane and chlorinated di-isopropyl ether. Some of the disadvantages and major economic drawbacks of the process which led to the wide acceptability of epoxidation processes are use of costly chlorine, production of weak calcium chloride byproduct, and corrosion problem due to chlorine handling.

Oxidation Route using peroxide Compounds:

In this process, propylene and peracetic acid (in ethyl acetate) which is produced by oxidation of acetaldehyde are reacted in a series of three specially designed reactors at 50-80 °C and 90-120 MPa pressure. The reaction products are fed to the stripper where a mixture of propylene and propylene oxide are obtained as top product while mixture of ethyl acetate and acetic acid is obtained as bottom product. Both mixtures are fed to two separated columns where separation of propylene oxide, ethyl acetate, acetic acid, and heavy end takes place.

Reaction

Peroxide from Acetaldehyde

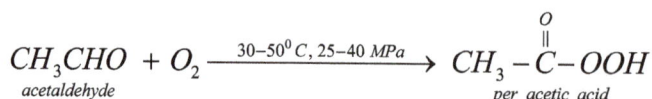

$$CH_3CHO + O_2 \xrightarrow{30-50^\circ C,\ 25-40\ MPa} CH_3 - \overset{\overset{\displaystyle O}{\displaystyle \|}}{C} - OOH$$

acetaldehyde　　　　　　　　　　　　　　　　per acetic acid

Oxidation of Propylene

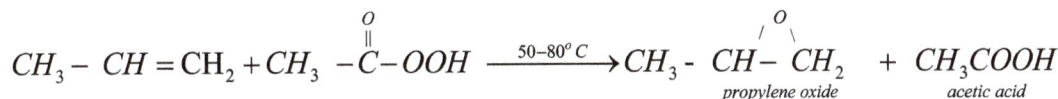

$$CH_3 - CH = CH_2 + CH_3 - \overset{O}{\overset{\|}{C}} - OOH \xrightarrow{50-80°C} CH_3 - \overset{O}{\overset{/\ \backslash}{CH-CH_2}} + CH_3COOH$$
$$\text{propylene oxide} \qquad \text{acetic acid}$$

Propylene Glycol

Propylene glycol is made by hydrolysis of propylene oxide. The process steps involve are:

Reaction Section: Hydrolysis of propylene oxide resulting in formation of mono propylene glycols(MPG). Small amount of di propylene glycol (DPG) and tri propylene glycol (TPG) s are also formed

Concentration Section: Concentration of glycol solution in multiple effect evaporator

Distillation Section: Separation of MPG,DIPG and TPG separated from MPG column. n series of distillation column where MPG is separated in first column.

Polyols

Polyols are made by polymerization of propylene oxide/ethylene oxide using an proprietary catalysed chain starter. The process consist of

- Raw material Preparation: Preparation of chain starter and addition in reactor along with EO/PO

- Reaction: Polymerisation using catalysed cahin extender

- Purification: Purification of raw polyol by neutralization

Isopropanol

Ever since its first commercial introduction in 1920 as one of the first petrochemicals, isopropyl alcohol has found wide use as a solvent and raw material for other chemical products like acetone, isopropyl acetate, glycerol, isopropyl and disopropyl amines, corrosion inhibitor di - soprpopyl ammonium nitrate, floatation agent isopropyl xanthate, isopropyl myristates etc.

Process Technology:

Two major processes for isopropanol manufacture are

- Esterification of propylene by sulphuric acid and hydrolysis

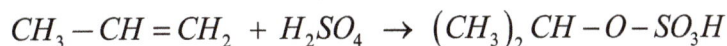

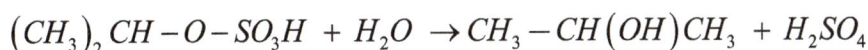

$$CH_3 - CH = CH_2 + H_2SO_4 \rightarrow (CH_3)_2 CH - O - SO_3H$$

$$(CH_3)_2 CH - O - SO_3H + H_2O \rightarrow CH_3 - CH(OH)CH_3 + H_2SO_4$$

- Direct catalytic hydration of propylene (vapor phase, liquid phase and mixed phase)

$$CH_3 - CH = CH_2 + H_2O \rightarrow CH_3 - CHOH - CH_3 \quad \Delta H_{298^0 K} = -51 \, KJ \, / \, mol$$

Although originally isopropyl alcohol was made by esterification of propylene and hydrolysis, problems of corrosion and a high heat requirement has led to the use of direct hydration process.

Direct Hydration of Propylene: In liquid phase hydration of propylene(Tokuyama Process) silico tungstate is used. The catalytic hydration process takes at 250-27°C at 200 atm pressure. Propylene conversion has been reported around 60-70%.

Butanols (N-Butanol and Iso-Butanol)

Various routes for making butanol are

- Acetaldehyde route

- Hydroformylation of propylene

- Oxidation of Butane

Condensation of Acetaldehyde: The process involves Aldolization, dehydration, hydrogenation

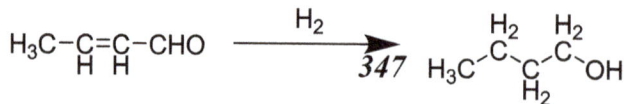

$$2 \, CH_3CHO \rightarrow CH_3CH(OH)CHO$$
$$\text{acetaldehyde}$$

$$CH_3CH(OH)CHO \rightarrow H_3C - \underset{H}{C} = \underset{H}{C} - CHO + H_2O$$

$$H_3C - \underset{H}{\overset{}{C}} = \underset{H}{\overset{}{C}} - CHO \quad \xrightarrow[347]{H_2} \quad H_3C - \overset{H_2}{\underset{}{C}} - \overset{H_2}{\underset{H_2}{C}} - \overset{H_2}{\underset{}{C}} - OH$$

Hydroformylation of Propylene

Butanol is manufactured from hydrogenation of n-butyraldehyde and iso-butyraldehyde mixture obtained by hydroformylation reaction of propylene and synthesis gas. Hydrogenation takes place at temperature 150-200 °C and 5-10 MPa pressure using copper or nickel catalyst. The butanols from the hydrogenation reactor go to a series of distillation columns for separation of light effluents and n-butanol and iso-butanol. About 88% of n-butanol and 12% iso-butanol are obtained.

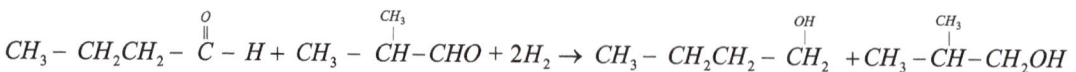

$$CH_3 - CH_2CH_2 - \overset{O}{\overset{\|}{C}} - H + CH_3 - \underset{CH_3}{\overset{}{C}}H - CHO + 2H_2 \rightarrow CH_3 - CH_2CH_2 - \underset{OH}{\overset{}{C}}H_2 + CH_3 - \underset{CH_3}{\overset{}{C}}H - CH_2OH$$

Butadiene [$CH_2=CH-CH=CH_2$]

Butadiene is one of the major petrochemicals with a wide range of uses as feedstock for production of a variety of synthetic rubbers and polymer resins, the bulk of which are related to styrene butadiene rubber (SBR), nitrile rubber, chloroprene rubber, polybutadiene rubber, and acrylonitrile butadiene styrene (ABS) resin. Another major use of butadiene is in the manufacture of adiponitrile which is a raw material for the production of Nylon 66. Global demand growth for butadiene is set to accelerate. Butadiene based synthetic rubbers are mainly used in the automotive industry. It is also widely used for manufacturing of engineering resins.

There are four major routes for production of butadiene:

- Steam cracking of naphtha

- Catalytic dehydrogenation of butenes

- Catalytic dehydrogenation of butanes

- Dehydrogenation-dehydration of ethanol (molasses route)

2-   ETHYL HEXANOL

Major use of 2-ethyl hexanol is in the manufacture of di -2-ethylhexylphthalate which is used as plasticiser for vinyl resins. Other application are in synthetic lubricants, antioxidants and antifoams. 2-Ethyl hexanol is made either by the oxo-synthesis or from acetaldehyde route by condensation and hydrogenation. . 2-Ethyl hexanol is also used in the manufacture of ethyl hexa acrylate. Ethyl hexaacylateprodices soft and tacky film with excellent low temperature flexibilities. Ethylhexanol also find application cable coating compositions, nitrocellulose lacquers, as softener in nitrile rubber compounds, in plastic compounds for water proof agents .

Propylene Route: In first step 4n-butyraldehyde is produced along with 1-isobutyraldehyde. 4n-butyraldehyde is further hydrogenated to 2-ethylhexanol

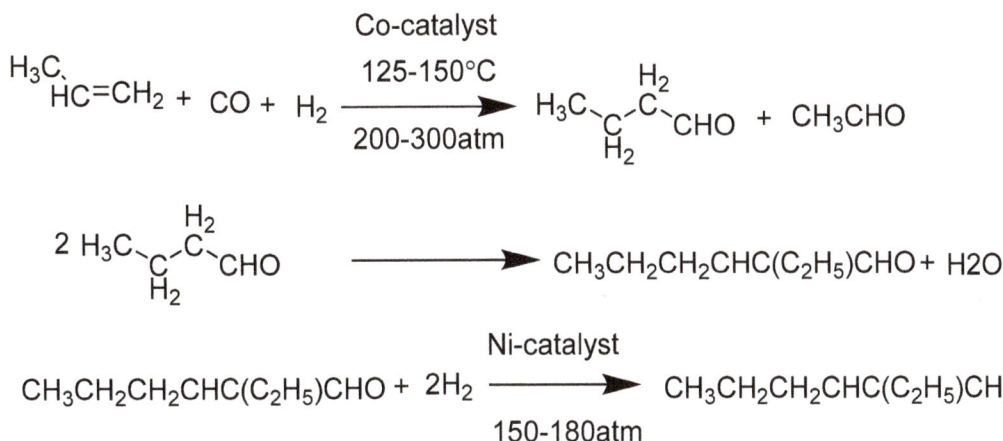

$$2\ H_3C-CH_2-CH_2-CHO \longrightarrow CH_3CH_2CH_2CHC(C_2H_5)CHO + H2O$$

$$CH_3CH_2CH_2CHC(C_2H_5)CHO + 2H_2 \xrightarrow[150-180atm]{Ni\text{-}catalyst} CH_3CH_2CH_2CHC(C_2H_5)CH_2OH$$

Propylene Carbonate [C$_3$H$_6$CO$_3$]

Propylene carbonate is prepared by reaction of propylene oxide and carbon dioxide in presence of ion-exchange resins

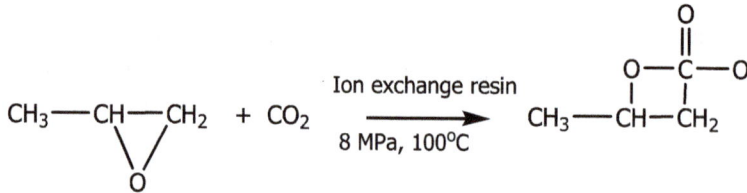

Uses: Propylene carbonate is used as special solvent. It is used in solvent extraction, plasticisers, organic synthesis, natural gas purification, and fiber spinning solvent.

Acrylic Acid

Acrylic acid is a versatile chemical which find application in the manufacture of glacial acrylic acid and acrylic esters (Acacrylates and metha acrylates), polyacrylic acid which is used in manufacture of super absorbent polymers, flocculants, detergents, paper chemicals and resin. SAP is used for water retention in infants diaper, adult in continence products and feminehygine products [Nandini Chemical Journal, July, 1999]. Various acrylic esters are methyl acylate, ethyl acrylate, butyl acylate, 2-ethyl hexyl acrylate.

Process Technology:

Various routes for making acrylic acid are

- Acetylene route
- Ethylene Oxide Route
- Ethylene Route
- Chlorination of Propianic acid
- Propylene route
- Formaldehyde and Acetic Acid Route

Amongst the above process propylene oxidation through acrolein is commonly used

Propylene Route: In this route, acrolein is made in first stage by oxidation of propylene in presence of mixed catalysts (prepared from oxides of bismuth, potassium, cobalt, and iron, nickel, tin, tellurium, tungsten, etc). In the second stage, acrolein is oxidised to acrylic acid in the presence of mixed oxides of molybdenum and vanadium at 250-280 °C in the presence of steam.

$$CH_2 = CH - CH_3 + O_2 \rightarrow CH_2 = CH - CHO$$
Acrolein

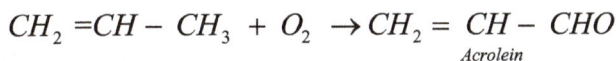

Aromatic Hydrocarbon

An aromatic hydrocarbon or arene (or sometimes aryl hydrocarbon) is a hydrocarbon with sigma bonds and delocalized pi electrons between carbon atoms forming a circle. In contrast, aliphatic hydrocarbons lack this delocalization. The term 'aromatic' was assigned before the physical mechanism determining aromaticity was discovered; the term was coined as such simply because many of the compounds have a sweet or pleasant odour. The configuration of six carbon atoms in aromatic compounds is known as a benzene ring, after the simplest possible such hydrocarbon, benzene. Aromatic hydrocarbons can be *monocyclic* (MAH) or *polycyclic* (PAH).

Some non-benzene-based compounds called heteroarenes, which follow Hückel's rule (for monocyclic rings: when the number of its π-electrons equals $4n + 2$, where $n = 0, 1, 2, 3,...$), are also called aromatic compounds. In these compounds, at least one carbon atom is replaced by one of the heteroatoms oxygen, nitrogen, or sulfur. Examples of non-benzene compounds with aromatic properties are furan, a heterocyclic compound with a five-membered ring that includes a single oxygen atom, and pyridine, a heterocyclic compound with a six-membered ring containing one nitrogen atom.

Benzene Ring Model

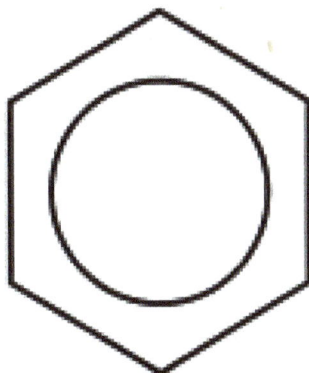

Benzene

Benzene, C_6H_6, is the simplest aromatic hydrocarbon, and it was the first one named as such. The nature of its bonding was first recognized by August Kekulé in the 19th century. Each carbon atom in the hexagonal cycle has four electrons to share. One goes to the hydrogen atom, and one each to the two neighbouring carbons. This leaves one electron to share with one of the same two neighbouring carbon atoms, thus creating a double bond with one carbon and leaving a single bond with the other, which is why the benzene molecule is drawn with alternating single and double bonds around the hexagon.

The structure is alternatively illustrated as a circle around the inside of the ring to show six electrons floating around in delocalized molecular orbitals the size of the ring itself. This depiction represents the equivalent nature of the six carbon–carbon bonds all of bond order 1.5; the equivalency is explained by resonance forms. The electrons are visualized as floating above and below the ring with the electromagnetic fields they generate acting to keep the ring flat.

General properties of aromatic hydrocarbons:

1. They display aromaticity

2. The carbon–hydrogen ratio is high

3. They burn with a strong sooty yellow flame because of the high carbon–hydrogen ratio

4. They undergo electrophilic substitution reactions and nucleophilic aromatic substitutions

The circle symbol for aromaticity was introduced by Sir Robert Robinson and his student James Armit in 1925 and popularized starting in 1959 by the Morrison & Boyd textbook on organic chemistry. The proper use of the symbol is debated; it is used to describe any cyclic π system in some publications, or only those π systems that obey Hückel's rule in others. Jensen argues that, in line with Robinson's original proposal, the use of the circle symbol should be limited to monocyclic 6 π-electron systems. In this way the circle symbol for a six-center six-electron bond can be compared to the Y symbol for a three-center two-electron bond.

Arene Synthesis

A reaction that forms an arene compound from an unsaturated or partially unsaturated cyclic precursor is simply called an aromatization. Many laboratory methods exist for the organic synthesis of arenes from non-arene precursors. Many methods rely on cycloaddition reactions. Alkyne trimerization describes the [2+2+2] cyclization of three alkynes, in the Dötz reaction an alkyne, carbon monoxide and a chromium carbene complex are the reactants. Diels–Alder reactions of alkynes with pyrone or cyclopentadienone with expulsion of carbon dioxide or carbon monoxide also form arene compounds. In Bergman cyclization the reactants are an enyne plus a hydrogen donor.

Another set of methods is the aromatization of cyclohexanes and other aliphatic rings: reagents are catalysts used in hydrogenation such as platinum, palladium and nickel (reverse hydrogenation), quinones and the elements sulfur and selenium.

Arene Reactions

Arenes are reactants in many organic reactions.

Aromatic Substitution

In aromatic substitution one substituent on the arene ring, usually hydrogen, is replaced by another substituent. The two main types are electrophilic aromatic substitution when the active reagent is an electrophile and nucleophilic aromatic substitution when the reagent is a nucleophile. In radical-nucleophilic aromatic substitution the active reagent is a radical. An example of electrophilic aromatic substitution is the nitration of salicylic acid:

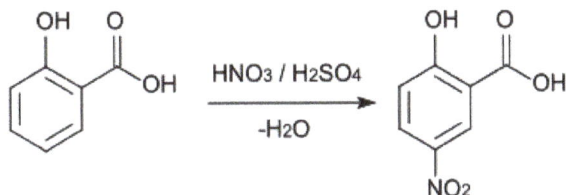

Coupling Reactions

In coupling reactions a metal catalyses a coupling between two formal radical fragments. Common coupling reactions with arenes result in the formation of new carbon–carbon bonds e.g., alkylarenes, vinyl arenes, biraryls, new carbon–nitrogen bonds (anilines) or new carbon–oxygen bonds (aryloxy compounds). An example is the direct arylation of perfluorobenzenes

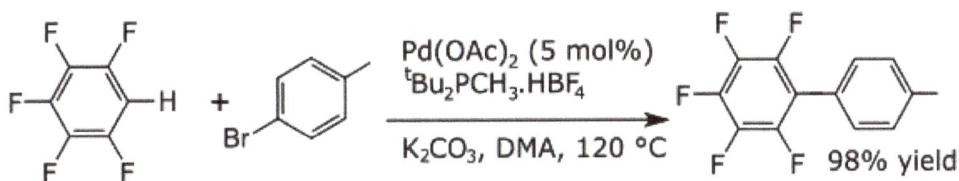

Hydrogenation

Hydrogenation of arenes create saturated rings. The compound 1-naphthol is completely reduced to a mixture of decalin-ol isomers.

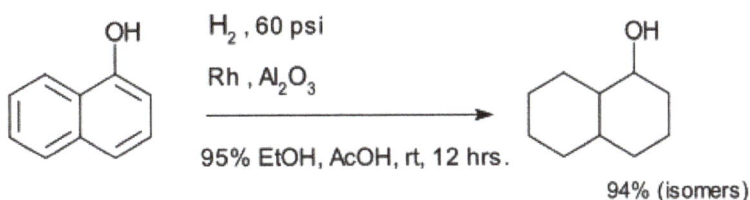

The compound resorcinol, hydrogenated with Raney nickel in presence of aqueous sodium hydroxide forms an enolate which is alkylated with methyl iodide to 2-methyl-1,3-cyclohexandione:

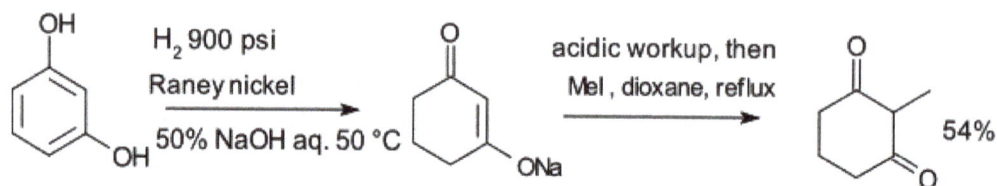

Cycloadditions

Cycloaddition reaction are not common. Unusual thermal Diels–Alder reactivity of arenes can be found in the Wagner-Jauregg reaction. Other photochemical cycloaddition reactions with alkenes occur through excimers.

Benzene and Derivatives of Benzene

Benzene derivatives have from one to six substituents attached to the central benzene core. Examples of benzene compounds with just one substituent are phenol, which carries a hydroxyl group, and toluene with a methyl group. When there is more than one substituent present on the ring, their spatial relationship becomes important for which the arene substitution patterns *ortho*, *meta*, and *para* are devised. For example, three isomers exist for cresol because the methyl group and the hydroxyl group can be placed next to each other (*ortho*), one position removed from each other (*meta*), or two positions removed from each other (*para*). Xylenol has two methyl groups in addition to the hydroxyl group, and, for this structure, 6 isomers exist.

- Representative arene compounds

Toluene	Ethylbenzene	p-Xylene
m-Xylene	Mesitylene	Durene

		OH
2-Phenylhexane	Biphenyl	Phenol
Aniline	Nitrobenzene	Benzoic acid
Aspirin	Paracetamol	Picric acid

The arene ring has an ability to stabilize charges. This is seen in, for example, phenol (C_6H_5–OH), which is acidic at the hydroxyl (OH), since a charge on this oxygen (alkoxide –O$^-$) is partially delocalized into the benzene ring.

Polycyclic Aromatic Hydrocarbons

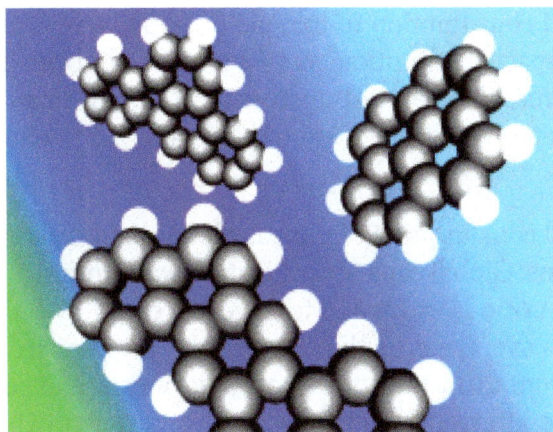

An illustration of typical polycyclic aromatic hydrocarbons. Clockwise from top left: benz(e)acephenanthrylene, pyrene and dibenz(ah)anthracene

Polycyclic aromatic hydrocarbons (PAHs) are aromatic hydrocarbons that consist of fused aromatic rings and do not contain heteroatoms or carry substituents. Naphthalene is the simplest example of a PAH. PAHs occur in oil, coal, and tar deposits, and are produced as byproducts of fuel burning (whether fossil fuel or biomass). As pollutants, they are of concern because some compounds have been identified as carcinogenic, mutagenic, and teratogenic. PAHs are also found in cooked foods. Studies have shown that high levels of PAHs are found, for example, in meat cooked at high temperatures such as grilling or barbecuing, and in smoked fish.

They are also found in the interstellar medium, in comets, and in meteorites and are a candidate molecule to act as a basis for the earliest forms of life. In graphene the PAH motif is extended to large 2D sheets.

Aromatics are backbone of organic chemical industries. Aromatic hydrocarbons especially benzene, toluene, xylene (BTX), and ethyl benzene are major feedstock for large number of intermediates which are used in the production of synthetic fibers, resins, synthetic rubber, explosives, pesticides, detergent, dyes, intermediates, etc. Styrene, linear alkyl benzene, and cumene are the major consumer of benzene. Product profile of aromatics is shown in Figure.

Global production and consumption of benzene in 2009 were between 36.4 and 36.6 million metric tons. Average global capacity utilization was 69.2percent in 2009, lower than in 2008. Global benzene consumption is estimated to have decreased by 3.8percent in 2009; however, it is expected to average growth of 4.7percent per year from 2009 to 2014, and 2.6percent per year from 2014 to 2019 [Petrochemical overview, SRI consulting]. Ethyul benzene/styrene monomer and cumene/phenol with demand shares of 52.6percent and 18.3percent respectively of the global benzene market respectively are the leading end-use segments for benzene in 2010. Other uses such as cyclohexane, nitrobenzene and LAB collectively consumed less than 15percent of global benzene demand in 2010.

Global production and consumption of toluene in 2009 were each almost 18.4 million metric tons. Global capacity utilization was 64percent in 2009. Toluene consumption estimated to have decreased by almost 2percent in 2009. Demand is expected to grow on average 3.6percent per year from 2009 to 2014, and 2.2 percent per year from 2014 to 2019.

Global production and consumption of mixed xylenes in 2009 were each approximately 41 million metric tons. Global capacity utilization was around 74percent in 2009. Xylenes consumption in 2009 is estimated to have increased by around 4.6percent from 2008; it is expected to average growth of 5.2percent per year from 2009 to 2014, and 3.3percent per year from 2014 to 2019. Operating rates are expected to drop in 2010 but gradually increase afterward [Petrochemical overview, SRI consulting]. Product profiles of major aromatics-BTX are given in Table.

Table: World Aromatic Petrochemicals Scenario

Product	Consumption 2002 ('000 tonnes)	Consumption Growth (in%)			Capacity 2002 ('000 tonnes)	Announced Capacity Due by 2012 ('000 tonnes)	Capacity Change Needed by 2012 2002 after Announcement	
		1997-2002	2002-2007	2007-2012			('000 tonnes)	(%)
Benzene	33,278	3.5	4.2	2.8	43,945	4,443	5,765	13
Toluene	16,688	1.9	3.5	2.4	24,642	1,184	900	4
Ethyl benzene	25,130	4.2	3.5	2.7	27,536	5,948	3,593	13
Styrene	22,188	3.9	3.5	2.7	23,742	6,087	6,647	28
Mixed Xylenes	29,187	5,7	5.1	4.1	36,784	6,640	9,775	27
Xylene	3,050	3.4	4.5	2.7	3,948	498	760	19
p-Xylene	18,701	6.2	5.8	4.6	20,605	7,359	7,320	36
Cumene	9,214	5.8	4.7	2.6	10,612	505	3,445	32
Phenol	7,166	5.2	4.4	1.7	7,843	1,262	2,078	26
Caprolactam	3,746	6.1	3.1	3.3	4,566	448	750	16
Terephthalic Acid	24,822	8.2	5.9	5.0	26,168	8,038	12,000	46
Phthalic Anhydride	3,488	3.5	4.1	2.5	4,374	358	700	16
Dimethyl Terephthalate	3,747	-3.7	1.5	0.5	4,710	53	65	1
TDI	1,383	3.9	4.1	3.7	1,809	932	0	0

Global production and consumption of p-xylene in 2009 were each approximately 27.6 million metric tons. Global capacity utilization was 83percent in 2009, unchanged from 2008. p-Xylene consumption is estimated to have increased by approximately 4.7percent in 2009; it is expected to average growth of 5.5percent per year from 2009 to 2014, and 3.5percent per year from 2014 to 2019[Petrochemical overview, SRI consulting].

Global production and consumption of o-xylene in 2009 were approximately 3.7 million metric tons. Global capacity utilization was 66percent in 2009. o-Xylene consumption is estimated to have increased slightly in 2009; it is forecast to average growth of 3.2percent per year from 2009 to 2014, slowing to 2.4percent per year from 2014 to 2019. Average global utilization rates are expected to remain in the 70s range throughout the forecast period [Petrochemical overview, SRI consulting].

Benzene →
- Naphthalene → Phthalic anhydride, insecticides, β-naphthol, moth balls, surface active agent, synthetic tanning agents, dyes, rubber chemicals, solvents, agricultural uses.
- Cumene → Phenol, Acetone → Adipic acid, aniline, Bisphenol-A, caprolactam, phenolic resin, pesticides, dyes, rubber chemical; Methyl isobutyl ketone, methyl methacrylate, solv methyl isobutyl carbinol.
- Linear alkyl benzene → LABS → Detergent
- Cyclohexane → Nylon 6, nylon 66, adipic acid.
- Ethyl benzene → Styrene → Polystyrene, SBR, ABS and SAN resins, styrenat polyester, acrylonitrile butadienes, styrene plastic protective coatings.
- Maleic Anhydride → Acids (fumaric, malic); lubricating oil additives; copolymers; agricultural chemicals,unsaturated polyester resin; 1,4-butanediol (Resins, polyurethane solvent, pharmaceuticals).
- Chloro benzene → Aniline, phenol, DDT, chloronitrobenzene, biphen
- Nitrobenzene → Aniline → Dyes and intermediates, rubber chemicals, drugs and pharmaceuticals, photographic chemicals, isocyanate.
- BHC → Pesticides

Toluene →
- Gasoline → Motor gasoline
- Nitrotoluene → Toluene diisocyanate → Polyurethane (Rigid foam, flexible foam, surface coatings)
- Trinitrotoluene → Explosives
- Benzoic acid → Caprolactam, pharmaceuticals and flavors, phthalates, terephthalic acid. / Phenol, sodium benzoate - food preservatives.
- Solvents
- p-cresol → Di-tert-butyl-p-cresol (antioxidants)

Xylenes →
- m-xylene → Isophthalic acid → Resins, unsaturated polyesters, plasticizers, othe esters.
- o-xylene → Phthalic anhydride → Plasticizers, polyester resins, alkyl resins, dyes and pigments, herbicides, isatoic anhydride, polyester polyols, phthalimide-fungicides.
- p-xylene → TPA, DMT → Polyester fibers, films, polyethylene, terephthalat

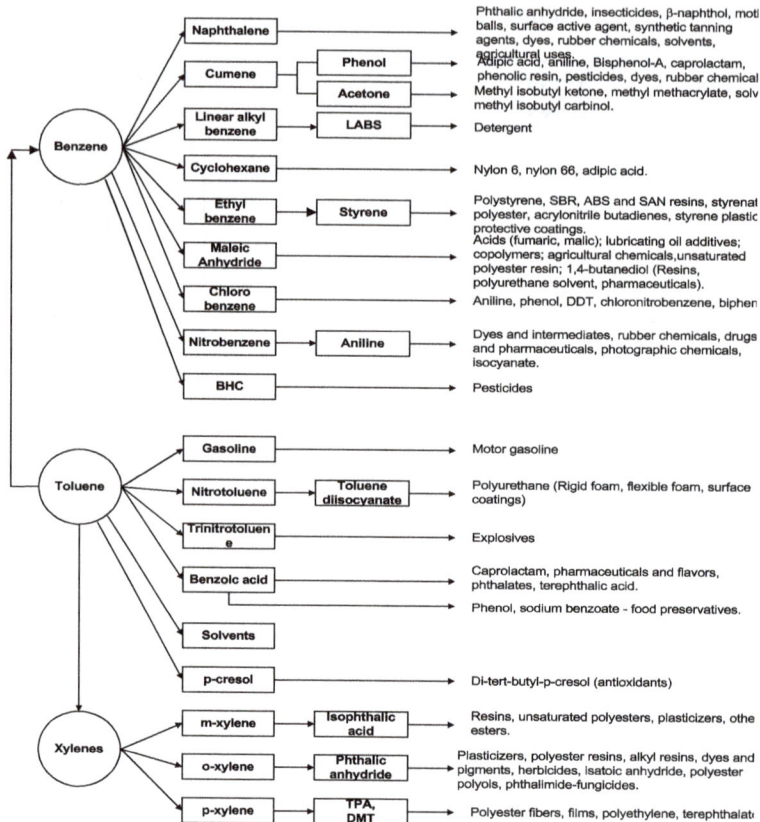

Product Profile of Aromatics

Ethyly Benzene and Styrene

Ethyl benzene and styrene are two important aromatics. Ethyl benzene is mainly used for making styrene. Styrene which finds application in synthetic rubber and polymer industry for the manufacture of SBR and polystyrene, ABS plastic.

Major route for styrene manufacture is dehydrogenation of ethyl benzene which is manufactured by alkylation of benzene. Styrene plant consists of two major units. The process involves:

- Production of ethylene either from molasses route or by naphtha/natural gas cracking

- Production of ethyl benzene by alkylation of benzene

- Dehydrogenation of ethyl benzene to styrene

Ethyl Benzene

Ethyl benzene is made by alkylation of benzene with ethylene. Ethylene can be produced from either from molasses route or naphtha/gas cracker. The convention alkylation catalysts are metal chlorides (BF_3, $AlCl_3$, etc) and mineral acids (HF, H_2SO_4). However, with development of zeolite, now the benzene alkylation I is done by using ZSM-5 catalysts using vapor-phase process [Mobil-badger Process] and liquid phase alkylation using MCM-22 azeolite proprietary catalyst based catalyst.

Vapour Phase Alkylation of Benzene: The process consists of vapor phase alkylation of benzene with ethylene using zeolite catalyst in a fixed bed catalytic distillation technology. Alkylation and distillation takes place in the alkylator. Unreacted ethylene, and benzene vapour are condensed and fed to the finishing reactor where the remaining alkylation is completed in the presence of a catalyst. The product stream goes to fractionating columns where ethyl benzene is separated from the higher ethylated benzene and heavy ends. Higher ethylated alkyl benzene is sent to the trans-alkylator where its trans-alkylated to produce additional ethyl benzene

Liquid Phase Alkylation Of Benzene With Ethylene Using MCM-22 Catalyst:In this process alkylation of ethylene takes place in a liquid filled alkylator reactor containing multiple fixed beds of MOBIL MCM-22 catalyst. During alkylation Ethyl benzene and small quantity of polyethylbenzene are formed which is converted to ethyl benzene using trans- alkylation catalyst. The product streams from alkylator and trans-alkylator are sent to various fractionating columns for separation of product ethyl benzene, polyethylbenzene, benzene, gases and heavy ends.

$$C_6H_6 + C_2H_4 \rightarrow C_6H_5CH_2CH_3$$

Styrene

Styrene is one of the most important monomers for the production of polymers, resins and rubber. The biggest consumer of styrene monomer is polystyrene, other major derivatives are expanded polystyrene, Styrene butadiene(SB) latex, SB rubber, styrene block co-polymers (eg: ABS, MBS, SBS) [SNOW: an innovative technology for styrene synthesis, hydrocarbon asia, 2007, p.42]

Styrene is made by catalytic dehydrogenation of ethyl benzene

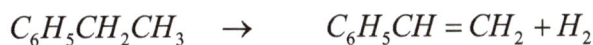

$$C_6H_5CH_2CH_3 \quad \rightarrow \quad C_6H_5CH = CH_2 + H_2$$

Styrene can be also directly recover from raw pyrolysis gasoline derived from cracking of naphtha, gas oils Using GT styrene process

Lumus/UOP EB One Process: Styrene is made by catalytic dehydrogenation of ethylbenzene in the presence of steam. In Lumus/UOP EB one process involves first alkylating benzene with ethylene followed by dehydrogenating the EB to form styrene. The benzene and recycled benzene are preheated the liquid phase reactor containing zeolite catalyst. The polyethylbenzene formed during alkylkation is fed to another reactor for transalkylating with benzene. Transalkylation reaction is isothermal and reversible in distillation section. The reactor effluent from both reactions is sent to the distillation section for separating ethyl benzene from polyethylbenzene.

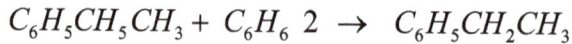

$$C_6H_6 + C_2H_4 \quad \rightarrow \quad C_6H_5CH_2CH_3$$

$$C_6H_5CH_2CH_3 + C_6H_6 \rightarrow 2C_6H_5CH_5CH_3$$

$$C_6H_5CH_5CH_3 + \ C_6H_6 \ 2 \ \rightarrow \ C_6H_5CH_2CH_3$$

Ethyl benzene and recycled ethyl benzene are then dehydrogenated to styrene in the presence of steam at high temperature (550-68°C) under vacuum in a multistage reactor

$$C_6H_5CH_2CH_3 \quad \rightarrow \quad C_6H_5CH = CH_2 + H_2$$

$$H_2 + 1/2O_2 \rightarrow H_2O$$

During dehydrogenation stages air or oxygen is introduced to partly oxidize the hydrogen to reheat the process gas and to remove the equilibrium constrain for dehydrogenation reaction. [HC,1999]. Reactor effluents are cooled to recover waste heat and condensed, uncondensed gases are used as fuel. The condensed product containing styrene is sent to distillation columns for separating styrene monomer, unconverted ethyl benzene is recycled.

Toluene is formed during the process which is recovered

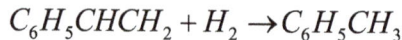

$$C_6H_5CHCH_2 + H_2 \rightarrow C_6H_5CH_3$$

$$C_6H_5CHCH_2 \rightarrow C_6H_5CH_3 + C$$

GT Styrene Process: Styrene can be also directly recovered from raw pyrolysis gasoline derived from cracking of naphtha, gas oils Using GT styrene process. Raw pyrolysis gasoline is fractionated into a heart cut C8 stream from which styrene is separated by extractive distillation.

Innovative SNOW Technology: The snow technology has been jointly developed by Snamprogetti and Dow represents a technological and economical breakthrough in styrene production and uses benzene and ethane as raw material which is dehydrogenated in the same reaction for EB dehydrogenation. SNOW reactor is rise type.

Phthalic Anhydride

Phthalic anhydride first became commercially important during the nineteenth century as an intermediate for dy stuff industry. However, now phthalic anhydride is largely used for the manufacture of plasticizers, alkyd resins, and unsaturated polyester resins where about 95percent of the phthalic anhydride production is consumed. With an aggregate installed capacity of 267,200-tpa across India, major PAN producers include IG Petrochemicals Ltd and Thirumalai chemicals Ltd. Consumption pattern of PAN is shown in Figure. List of the phthalic anhydride manufacturer in India is given in Table.

Consumption pattern of PAN

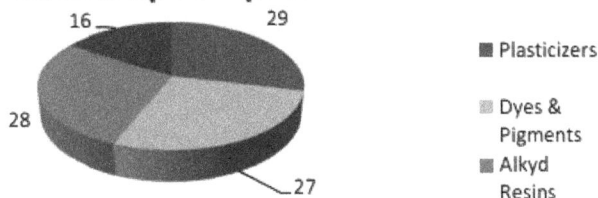

Consumption pattern of PAN

Table: Phathalic Anhydride Manufacturer in India

Company	Location	Installed Capacity (TPA)
I.G. Petrochemicals Ltd.	Taloja, Maharashtra	120,000
Thirumalai Chemicals Ltd.	Ranipet, Tamil Nadu	100,000
Asian Paints Ltd.	Ankleshwar, Gujarat	25,200
Mysore Petrochemicals Ltd.	Raichur, Karnataka	12,000
S.I. Group Ltd.	Thane, Maharashtra	10,000
Total		267,200

The basic raw material for the manufacture of phthalic anhydride is naphthalene and o-xylene. Phthalic anhydride was manufactured from naphthalene. With the availability of large amount of o-xylene as a byproduct during p-xylene production, now phthalic anhydride is made from o- xylene. Both vapor phase and liquid phase oxidation of o-xylene are available.

Phthalic anhydride is produced by oxidation of naphthalene in the gas phase using vanadium pentoxide catalyst supported on silica or silicon carbide promoted with various other metal oxides, e.g. titanium oxide (wire) in either a fixed bed multiple reactors or fluidized bed reactor

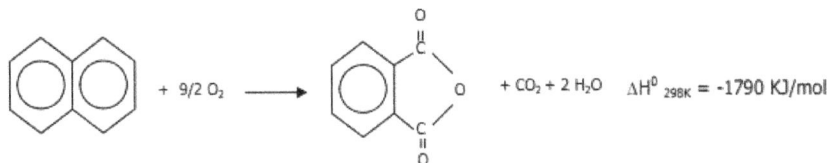

Production of phthalic anhydride from o-xylene is similar to naphthalene route. Catalytic oxidation of o-xylene is done either in fixed bed catalytic reactor having multi tube or fluidised bed reactor in the presence of vanadium pentoxide and titanium oxide catalyst.

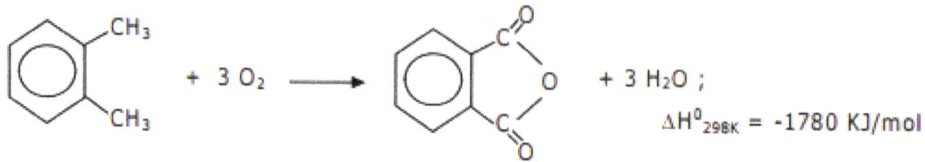

$$+ 3\, O_2 \longrightarrow \qquad + 3\, H_2O\ ;$$

$$\Delta H^0_{298K} = -1780\ KJ/mol$$

Cumene

Cumene is made by alkylkating benzene with propylene using zeolite catalyst. Following three major processes are available

Catalytic Distillation Technology: The process uses a specially formulated zeolite alkylation catalyst in a proprietary catalytic distillation (CD) process and a trans-alkylator reactor using zeolite catalyst. In CD column combines both reaction and fractionation takes place.

Liquid Phase Q-max Process: In this process, cumene is produced by liquid phase alkylation of benzene with propylene in presence of azeolite catalyst.

Cumene by Mobil Badger Process: The process produces cumene from benzene and any grade of propylene using a new generation of zeolite catalysts from Exxon m Mobil. The process includes a fixed bed alkylation reactor and a fixed bed trans-alkylation reactor and distillation section.

Phenol

According to SRI consulting report 2010 global production and consumption of phenol were both around 8.0 million tones with global capacity utilization of 77percent. Phenol consumption is expected to average growth of 5.1percent per year from 2009 to 2014 and around 2.5percent from 2014-19. Phenol is consumed mainly for production of bisphenol A and phenolic resins which accounted for 42percent and 28percent respectively of total phenol consumption in 2009. Various routes for Phenol:

- Phenol from Cumene
- Phenol from Benzoic acid
- Phenol from chlorobenzene
- Benzene Sulphonation

With the availability of propylene now phenol is made by cumene route with added advantage of acetone as by product

Aniline

The process of aniline manufacture involves two stages: Company which produce aniline is given in Table.

Table: Company-Wise Production of Aniline

Company	Years	Production	Sales Quantity	Sales Value
Gujarat Narmada Valley Fertilizers Ltd.	2008-09	27077	27090	1865
	2009-10	33848	33825	2167
	2010-11	39896	-	-
Hindustan Organic Chemicals Ltd.	2009-10	5538	5231	309
	2010-11	1833	1826	135

Nitrobenzene Route:

- Nitration of benzene with nitric acid

- Hydrogenation of nitrobenzene to aniline

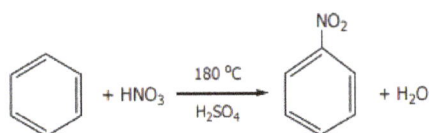

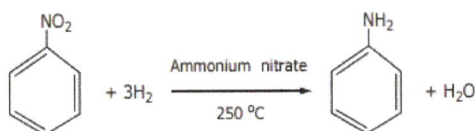

Ammonylysis of Chlorobenzene

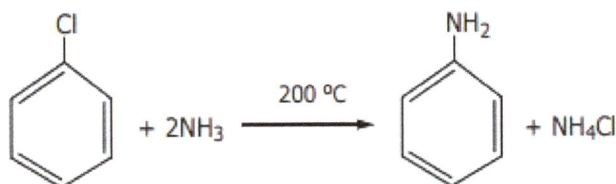

Ammonolysis of Phenol

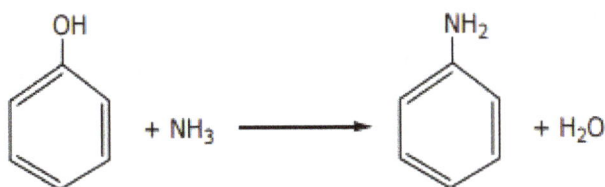

Benzoic Acid (C_6H_5COOH)

Benzoic acid is the simplest member of the aromatic carboxylic acid. Benzoic acid, which is used in the manufacture of caprolactam, phenol, terephthalic acid and used as mordant, is manufactured by liquid phase catalytic oxidation of toluene in presence of cobalt acetate at 165 °C and 11.2 atm pressure. Major processing steps in the manufacture of benzoic acid consist of:

- Catalytic liquid phase air oxidation of toluene

- Stripping of unreacted toluene and light end precursors from the benzoic acid for recycle

- Distillation to recover benzoic acid as a pure overhead product

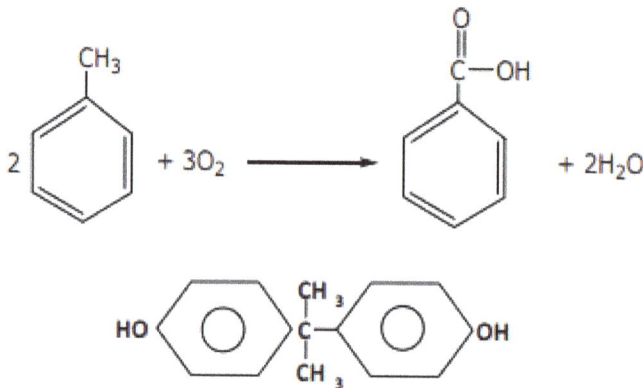

Bisphenol

Bisphenol is an important building block and its measure use is in the manufacture of polycarbonate plastic and epoxy resins. Other uses include in flame retardants, unsaturated polyester resin and polyacrylate, polyetherimide and polysulphone resin.

India and Global Demand of Bisphenol [Chemical Business, 2012]

Demand of bisphenol in India during 2010-11 was 30,000 tonnes per annum Global installed capacity: around 5.2 million tones

Global demand around 4.2 million tones

Global growth rate in demand 5 to 6 percent

Polycarbonate resin are the largest and fast growing BPA market, consuming 60percent of the global production.

Process Technology

Various process technologies available for manufacture of bisphenol are:

- Condensation of phenol with acetone

- Condensation of phenol with alkenyl phenol

- Condensation of phenol with ethylene and acetylenes

- Condensation of phenol with alkyl benzene

Bisphenol From Phenol And Acetone: Bisphenol is synthesized by a condensation reaction between phenol and acetone using proprietary cation exchange resin-base catalyst (4PET) in a packed bed reactor. The catalyst has higher acetone conversion, higher BPA selectivity and longer life. Reactor effluents are process in series of distillation column for separation of product bisphenol, unreacted acetone, water, phenol. Phenol and acetone are recycled. Bisphenol is purified by crystallization where bisphenol crystals are separated from the impurities. Although the impurities are removed with mother liquor, however two stage crystallistion can lower the impurities captured in the crystal. Bisphenol is sent to prilling tower to get final bisphenol in the form of spherical prill.

References

- Mick, Jason (3 March 2010). "Why Let it go to Waste? Enerkem Leaps Ahead With Trash-to-Gas Plans". DailyTech. Retrieved 22 February 2016

- Boehman, André L.; Le Corre, Olivier (2008). "Combustion of Syngas in Internal Combustion Engines". Combustion Science and Technology. Taylor & Francis. 180 (6): 1193–1206. doi:10.1080/00102200801963417

- "Syngas composition". National Energy Technology Laboratory, U.S. Department of Energy. Retrieved 7 May 2015

- "Syngas Production Using a Biomass Gasification Process". University of Minnesota. Retrieved 22 February 2016

- Carey, R.; Gomezplata, A.; Sarich, A. (January 1983). "An overview into submarine CO2 scrubber development". Ocean Engineering. 10 (4): 227–233. doi:10.1016/0029-8018(83)90010-0

- Matthew L. Wald (April 10, 2013). "New Solar Process Gets More Out of Natural Gas". The New York Times. Retrieved April 11, 2013

- Frances White. "A solar booster shot for natural gas power plants". Pacific Northwest National Laboratory. Retrieved April 12, 2013

- D'Alessio, L.; Paolucci, M. (1989). "Energetic aspects of the syngas production by solar energy: Reforming of methane and carbon gasification". Solar & Wind Technology. Elsevier. 6 (2): 101–104. doi:10.1016/0741-983X(89)90018-0

- "Audi in new e-fuels project: synthetic diesel from water, air-captured CO2 and green electricity; "Blue Crude"". Green Car Congress. 14 November 2014. Retrieved 29 April 2015

- Emmanuel O. Oluyede. "FUNDAMENTAL IMPACT OF FIRING SYNGAS IN GAS TURBINES". Clemson/EPRI. Retrieved 2016-06-13

Permissions

All chapters in this book are published with permission under the Creative Commons Attribution Share Alike License or equivalent. Every chapter published in this book has been scrutinized by our experts. Their significance has been extensively debated. The topics covered herein carry significant information for a comprehensive understanding. They may even be implemented as practical applications or may be referred to as a beginning point for further studies.

We would like to thank the editorial team for lending their expertise to make the book truly unique. They have played a crucial role in the development of this book. Without their invaluable contributions this book wouldn't have been possible. They have made vital efforts to compile up to date information on the varied aspects of this subject to make this book a valuable addition to the collection of many professionals and students.

This book was conceptualized with the vision of imparting up-to-date and integrated information in this field. To ensure the same, a matchless editorial board was set up. Every individual on the board went through rigorous rounds of assessment to prove their worth. After which they invested a large part of their time researching and compiling the most relevant data for our readers.

The editorial board has been involved in producing this book since its inception. They have spent rigorous hours researching and exploring the diverse topics which have resulted in the successful publishing of this book. They have passed on their knowledge of decades through this book. To expedite this challenging task, the publisher supported the team at every step. A small team of assistant editors was also appointed to further simplify the editing procedure and attain best results for the readers.

Apart from the editorial board, the designing team has also invested a significant amount of their time in understanding the subject and creating the most relevant covers. They scrutinized every image to scout for the most suitable representation of the subject and create an appropriate cover for the book.

The publishing team has been an ardent support to the editorial, designing and production team. Their endless efforts to recruit the best for this project, has resulted in the accomplishment of this book. They are a veteran in the field of academics and their pool of knowledge is as vast as their experience in printing. Their expertise and guidance has proved useful at every step. Their uncompromising quality standards have made this book an exceptional effort. Their encouragement from time to time has been an inspiration for everyone.

The publisher and the editorial board hope that this book will prove to be a valuable piece of knowledge for students, practitioners and scholars across the globe.

Index

www.ingramcontent.com/pod-product-compliance
Lightning Source LLC
Chambersburg PA
CBHW061954190326
41458CB00009B/2867